AF540441

CRYOGENICS

APPLIED SCIENCE SERIES

A TEXT BOOK OF
CRYOGENICS

Prof. Velery V. Kostiouk
Corresponding Member
Russian Academy of Science

Editor

Dr. Digumarti Bhaskara Rao
Secretary
Academy of Communication Culture
Education Science and Service
1-22-10 Srinivasa Nagar
Guntur-522006 A.P.

DISCOVERY PUBLISHING HOUSE
NEW DELHI-110002

Reprinted - 2019

First Published - 2003

ISBN: 978-81-7141-642-4

A Text Book of Cryogenics

Published by:

DISCOVERY PUBLISHING HOUSE PVT. LTD.
4383/4B, Ansari Road, Darya Ganj
New Delhi-110 002 (India)
Phone: +91-11-23279245, 23253475; 43596065
E-mail: discoverybooksindia@gmail.com
discoverypublishinghouse@gmail.com
web: www.discoverypublishinggroup.com

Printed at:
Infinity Imaging Systems
Delhi

Contents

Editor's ...

Cryogenics, a new branch of applied science, in having a very strong impact on technology and industrial production. This scientific-technological discipline has been introduced in curricula of many engineering and technological colleges and universities all over the world.

This book covers a wide variety of concepts concerned to cryogenics. The material on cryogenics is meant for students, engineers, technologists, and researchers studying and working in different fields of science and technology.

This book-cryogenics-is a part of my Series-Applied Science-of books for ready use.

Dr. D. Bhaskara Rao

Preface

Cryogenics is a new branch of science. The questions considered here are urgent for power engineering, electronics, nuclear and rocket space technology, food industry, scientific researches, etc.

The content is prepared on various sections of cryogenics: properties of cryogenic products and solid bodies at low temperatures, methods of internal cooling, heat transfer at low temperatures, gas liquefaction and separation, microcryogenic systems, storage and transportation of cryogenic fluids, cryogenic systems for scientific research.

This book is useful for technical courses.

Prof. V.V. Kostiouk

Acknowledgements

The editor is thankful to the Director, UNESCO-ROSTE, Venice, Italy for giving permission to reproduce the material from the UNESCO's publication entitled "Introduction to Cryogenics" prepared by Prof. V. V. Kostiouk as a part of Teaching Modules Series in Science and Technology, UNESCO-ROSTE.

The editor is also thankful to Prof. Kostiouk and Ms. Rosanna Santesso.

1

PROPERTIES OF CRYOGENIC PRODUCTS AND THEIR USE

Because of the extensive use of cryogenic temperatures in man's various branches of economic and scientific ctivities, the problems of effective applications of cryogenic products on an industrial scale are significant achievement and use of cryogenic temperatures (below 120 K) exploits a number of substances: helium, hydrogen, carbon oxide, fluorine, argon, oxygen, nitrogen, neon, and so on. The reason is that these substances provide cryogenic temperatures by using the simplest and most effective methods.

Nitrogen, oxygen, argon, neon, krypton, xenon are the main products separated from air, and are extracted.

1.1. Properties of cryogenic products

The most important thermodynamic states of cryogenic products are:

- the triple point;
- the boiling point at atmospheric pressure p_0;
- the critical point.

Parameters of the most important thermodynamic states of cryogenic products are given in Table 1.1.

Here T_{so} is saturation temperature of liquid at atmospheric pressure p_0; T_{tp} and p_{tp} are state parameters of the triple point; T_c and p_c are parameters of the critical point.

Table 1.1. Parameters of the most important thermodynamic states of cryogenic products.

Substance	T_{so} K	T_{tp} K	p_{tp} kPa	T_c K	p_c kPa
Helium	4.2	-	-	5.2	203
Hydrogen	20.4	13.9	7.2	33.2	1316
Neon	27.1	24.5	43.3	44.4	2653
Nitrogen	77.4	63.2	12.5	126.2	3396
Fluorine	85.1	53.5	0.0025	172.1	5680
Argon	87.0	83.8	68.7	150.7	4800
Oxygen	90.2	54.4	0.12	154.8	5040
Methane	111.7	90.7	11.72	190.8	4626
Krypton	119.8	115.8	73.1	210.1	5400
Xenon	165.0	161.4	81.6	290.0	5990
Ammonia	239.8	195.4	6.076	405.6	11297

1.1.1. Nitrogen

Two stable nitrogen isotopes with mass numbers 14 and 15 are known; these are in the proportion 10000 : 37, i.e. the volume fraction of N^{15} totals just 0.37%. In liquid state, nitrogen is transparent and colorless. Nitrogen ice has a greater density compared with the liquid and sinks in it. Values of latent heat of evaporation and melting for nitrogen are an order less than for water while the liquid nitrogen density is only 15% less than that for water. In solid nitrogen, at a temperature of about 35.6 K, an allotropic state change is observed, accompanied by increasing heat capacity. The transition heat amounts to about 8.2 $kJ \cdot kg^{-1}$. Nitrogen is inert, non-toxic, and has no magnetic properties or flavor. Nevertheless, manipulating liquid nitrogen requires some care because of the danger not only of a cold burn but also of explosion of gases condensing in liquid nitrogen; therefore, in particular, prolonged a contact between liquid nitrogen and atmospheric air, and hydrocarbons should be avoided.

1.1.2. Oxygen and ozone

Three stable oxygen isotopes with mass numbers 16, 17 and 18 in proportion 10000 : 4 : 20 are present in oxygen, i.e. the volume fraction of the O^{17} isotope is 0.04% and of the O^{18} isotope 0.2%. In liquid and solid oxygen, a very small amount of the polymer O_4 is formed; it is thought that this assigns a light-blue color to condensed oxygen. Despite the fact that the oxygen boiling point is almost 13 K higher than that for nitrogen, the oxygen solidification point is on the contrary almost 9 K below the nitrogen value. The liquid oxygen density is greater than the values for nitrogen and water. Oxygen ice sinks in the liquid. In solid oxygen, at temperatures of about 44 and 24 K, allotropic state changes are observed. At a temperature of 44 K, the phase transition heat is greater than the melting and is approximately equal to 23 $kJ \cdot kg^{-1}$. The transition heat at 24 K equals a total of 2.94 $kJ \cdot kg^{-1}$.

Unlike other gases, oxygen is paramagnetic. Some gaseous oxygen compounds also possess magnetic properties, e.g. nitrogen oxide NO is paramagnetic. By varying the susceptibility, it is possible to determine the oxygen concentration change in a diamagnetic gas mixture.

Oxygen is a chemically active gas and is a strong oxidant. It is especially dangerous when mixed with hydrocarbons; detonating admixtures of crystal hydrocarbons present in liquid oxygen have frequently been the cause of explosions and accidents. Contact betveen liquid oxygen and oils, fats, coal, wood, tissues, asphalt is prohibited. In working with oxygen, strict observation of the safery rules is essential. In developing and designing oxygen equipment, careful attention should be paid to choosing structural and working materials (insulation, adsorbents, supports, etc.). Oxygen is toxic; its toxicity limit is near its 60% concentration at normal atmospheric pressure. Pure oxygen may be used for breathing at decreased pressures.

Ozone. Three oxygen atoms are combined in an ozone molecule. Ozone is toxic, has a strong smell and is a very active oxidant. With more than 10^{-5} % of volume ozone present in the air, poisoning is possible.

The normal boiling point of ozone is 161.25 K while the triple point temperature is 80.65 K; pure liquid ozone may, therefore be supercooled by oxygen, nitrogen or air. Nowever, even under strong supercooling, a small excitation may result in ozone detonation. Ozone easily detonates not only in liquid phase but also in solid state when strongly disturbed; therefore, practical use of ozone remains difficult.

1.1.3. Inert gases

Argon. This is the third component, by volume content, of atmospheric air. It has three stable isotopes with mass numbers 40, 38 and 36 in the proportion 10000 : 6.3 : 33.8. The temperature difference betveen boiling and melting points amounts to just 3.44 K . Argon is distinguished by a relatively small mass heat capacity both in liquid and gaseous state. It is characteristic that the ratio of the liquid heat of evaporation to its supercooling heat by one degree (phase conversion criterion) is almost 1.5 time higher for argon than for nitrogen; argon is therefore easily evacuated by cryogenic pumping methods. Argon is inert and non-toxic. Liquid argon is colorless and transparent. Liquid argon has a greater density than oxygen. Argon ice sinks in the liquid.

Neon. Three stable neon isotopes with mass numbers of 20, 21 and 22 are present in neon extracted from air in the proportion 10000 : 28 : 971, i.e. the volume fraction of Ne^{22} emounts to 9.71%. Neon is inert and nontoxic. Possessing a low boiling point (27.1 K) and comparatively high density, neon is a promising oxidant, especially when used in turbo-machines. Liquid neon is transparent and colorless. Neon ice has a greater density than the liquid and sinks in it.

Krypton. This is an inert, colorless gas with no flavor. The liquid is also transparent and colorless. The liquid density is 2413 $kg \cdot m^{-3}$. The normal boiling point is 119.8 K; the melting point is 116.15 K. The evaporation heat amounts to 115 $kJ \cdot kg^{-1}$. Krypton extracted from air contains six stable krypton isotopes with mass numbers of 78 (0.354%); 80 (2.27%); 82 (11.56%); 83 (11.55%); 84 (57.02%); 86 (17.37%).

Xenon. This is inert and non-toxic. There exist nine stable krypton isotopes in the air. The largest volume fraction accounts for isotopes with mass numbers of 129 (26.44%); 130 (4.08%); 131 (21.18%); 132 (26.89%); 134 (10.44%); 136 (8.7%). A liquid xenon density is 3057 $kg \cdot m^{-3}$, i.e. more than three times exceeds that of water. Liquid xenon is transparent and boils at $T = 165.05$ K at normal atmospheric pressure. The melting point is 161.25 K. The evaporation heat is 96.3 $kJ \cdot kg^{-1}$.

1.1.4. Fluorine

This has one stable isotope with a mass number of 19. It is very chemically active and extremely toxic. It possesses a pungent smell. Liquid fluorine is light yellow. Fluid fluorine has a density 1.3 times higher than for oxygen. Its boiling point is only 5.64 K below the oxygen boiling point. Fluorine ice is also yellow, and sinks in the liquid.

In a supercooled solid phase at $T \approx 45.6$ K, an allotropic state change is observed, fluorine ice acquiring a white color. Fluorine is the most active oxidant. It reacts with all organic materials, water and with almost all inorganic substances. Even ceramic powder-like compounds may be burnt in the fluorine medium.

On the surface of the majority of metals while in contact with fluorine, a thin protective film is formed, which hinders the development of a reaction. For this reason, at low and moderate temperatures (up to approx. 470 K) corrosion-resistant steels, nickel, monel, brass, copper, and aluminum may serve as structural materials when using solid silver solder welding and soldering and passivating working surfaces. Sealing screw junctions allow use of a thin elastic teflon strip.

Equipment must be carefully washed, degreased, dehydrated, and blown-through by dry nitrogen.

Nevertheless, when filling the equipment with fluorine, especially on first contact with fluorine, it is necessary to be ready to take emergency safety measures. Because of the high toxicity, wide use of fluorine is not likely to occur.

1.1.5. Hydrogen

There exist three hydrogen isotopes: protium H with a mass number of 1; deuterium D with a mass number of 2; tritium T with a mass number of 3. An atomic nucleus of protium contains one proton; of deuterium, one proton and one neutron; of tritium, one proton and two neutrons. All isotopes have one electron. Protium and deuterium are stable. The half-life of tritium is 12.262 years. He^3 is a decay product.

Irrespective of the production method, hydrogen contains up to 99.987% of protium atoms. The deuterium atom content varies from 0.013 to 0.016%. As the hydrogen molecule is diatomic, the absolute majority of deuterium atoms are combined into molecules with protium atoms, thus forming so-called hydrogen deuteride HD, whose content amounts to 0.026 ÷ 0.032%. The boiling point of HD is approx. 22.13 K; it can, therefore be extracted by the low-temperature rectification method.

There exist two modifications of hydrogen: ortho-hydrogen ($o - H_2$) and para-hydrogen ($p - H_2$). These are distinguished by the direction of rotation of nuclei, i.e. by nuclear spin (Fig. 1.1).

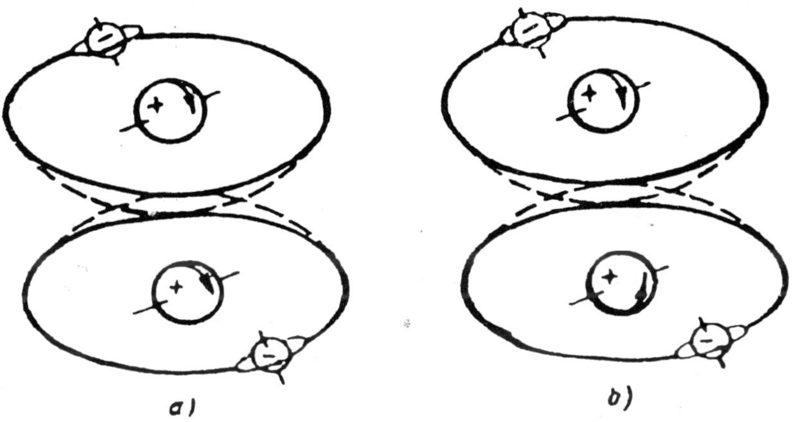

Fig. 1.1. A scheme of ortho-hydrogen (a) and para-hydrogen (b) molecules.

The equilibrium composition of ortho and para modifications varies with temperature. At temperatures of 100 ÷ 300 K, hydrogen is a mixture of 75% of the ortho modification (nuclear spin in one direction) and 25% of the para modification (nuclear spin in the opposite direction spins); such hydrogen is called normal. With decreasing temperature, the contribution of the para modification increases.

At equilibrium, liquid hydrogen contains 99.8% of the para modification and 0.2% of the ortho modification, i.e. equilibrium liquid hydrogen is practically para hydrogen. The ortho-para transformation is called ortho-para transition or ortho-para conversion. In gaseous state, ortho-para conversion is possible only in the presence of catalysts. In liquids phase, it proceeds spontaneously, very slowly.

For example, liquified normal hydrogen first has the composition of initial gaseous hydrogen; the fraction of para hydrogen molecules grows in time and amounts in 100 h to

0.595, giving in 1000 hours approx. 0.92. The conversion process is accompanied by heat release. At low temperatures (15 K < T < 70 K) the heat of orthotransition is approximately the same and is 706 $kJ \cdot kg^{-1}$. With rising temperature, it rapidly decreases, and for example at 200 K is only 219 $kJ \cdot kg^{-1}$.

Hence, it is seen that the heat release on conversion in liquefied normal hydrogen is greater than the evaporation heat (447 $kJ \cdot kg^{-1}$); therefore, even a perfectly insulated vessel, normal liquid hydrogen may evaporate during ortho-para transformation. Liquid losses in this case are approx. 18% for the first 24 hours, and in 100 hours less than 60% of liquid remains.

To reduce losses of liquid hydrogen when stored, with the latter liquefied, it should be insisted that accelerated conversion be made in the presence of solid catalysts. Mostly, conversion is arranged at several temperature levels: e.g. at temperatures of 65 ÷ 75 and 20 K.

Acceleration of the conversion in the presence of catalysts is attributed to the fact that on a catalyst surface steady adsorption occurs: ortho hydrogen molecules incident on the catalyst surface leave it para modified. Catalysts may be represented by various solid substances: activated coal, metal oxides, iron, nickel, chromium, manganese hydroxides. The structure of catalysts is refined, with a grain size about 1 mm. The conversion process in liquid normal hydrogen may also be accelerated by ultrasound. However, this method is not used in practice. If liquid para hydrogen evaporates and heats up, then for a long time it will remain para modified even at temperatures above 300 K. To return to the equilibrium composition, hydrogen is heated to 1000 K in the presence of catalysts (nickel, tungsten, platinum).

The hydrogen state surface has no specific features, but nevertheless hydrogen possesses a number of unique properties.

It is the lightest gas, its density for all states is smaller than that of other substances; therefore, the specific mass heat capacity in any aggregation state is highest.

The hydrogen molecule velocity is greater than that of other gases, and is therefore characterized by the highest diffusion velocity. In gaseous state, hydrogen is distinguished by the highest thermal conductivity and lowest viscosity. Hydrogen ice density, as well as the density of other cryogenic substances, is greater than that of liquid, in which it sinks. There exist differences in the physical properties of ortho and para modifications. In this situation, the differences in the values of the density, evaporation heat, melting heat, thermal conductivity of liquid, and sonic velocity are not great and nead not be taken into account in engineering calculations. The largest differences are typical only of heat capacity and, hence, of thermal conductivity and over a temperature range 80 ÷ 250 K, i.e. when a contribution of the rotational motion to heat capacity degenerates. Over this temperature range, the values of heat capacity and thermal conductivity of para hydrogen may exceed those of normal nydrogen by up to 20%.

Helium solubility in liquid hydrogen at a pressure up to 0.3 MPa is close to zero. As a rezult, gaseous and colder helium may by used, e.g. for liquid hydrogen supercooling, thus causing bubbling.

Hydrogen burns in the presence of air or oxygen with a wear-blue or almost colorless (if no impurities) flame with a high propagation velocity (2.7 $m \cdot s^{-1}$) and low emissivity. Ignition of hydrogen mixed with air and oxygen depends on hydrogen concentration. For example, for hydrogen mixed with air at atmospheric pressure and at 293 K the ignition range is 4 to 75% of hydrogen, and as regards possibilits of explosion, the most dangerous range of hydrogen concentrations is 18 ÷ 65%. Gaseous hydrogen is, to a considerable extent, a fire hazard, and work with it requires care. Detonation (existence of a stable detonation wave) originates mainly with ignition in finite volumes (apparatus, tubes, motors, rooms, etc.) and is seldom observed in open space.

With liquid hydrogen used, in addition, there appears danger of explosion because of possible formation of detonating condensed mixtures: liquid hydrogen (solid oxygen and liquid hydrogen) solid oxygen - enriched air.

Special attention is therefore paid to liquid hydrogen purity and to control of its composition.

1.2. Use of cryogenic products

Not all the above-mentioned substances find wide use at low temperatures. Their possible use at cryogenic temperatures is affected by two main factors: use of substances in different branches of technology and as cryogenic agents. Small content in the atmosphere and high cost of some (neon), limited use and toxicity of others (fluorine, carbon oxide), wide use and relative accessibility of still others (oxygen, hydrogen, methane, argon, nitrogen), as well as exclusive properties (nitrogen, helium) when used as cryogenic agents have resulted in the fact that at present, the substances methane, argon, oxygen, nitrogen, hydrogen and helium, are used at cryogenic temperatures on an industrial scale.

Methane and hydrogen are used as chemical raw material and fuel; argon, as an inert substance; nitrogen, as an inert substance, chemical raw and cryogenic agent; oxygen, as an active oxidant; helium, as the lowest temperature cryogenic agent.

Use of the substances mentioned in various branches of technology is often connected with their use in liquid state. This is dictated by several factors which specify the advantages of a liquid state as against a gaseous one. First, liquid is 800 times denser than a gas under normal conditions. This allows considerable reduction of volume and mass of storing and shipping containers, as well as making it technically possible to accumulate, store and dispatch large quantities of working products to users. This is especially important in cases where chemical properties of products are used.

Storing, transporting, creating reserves, providing hydrogen, oxygen, argon, methane in liquid state for a particular delivery date, with the latter gasified during the delivery, are more profitable and sometimes the only possible ways of using the above-mentioned substances in a certain branch of technology.

Thus, only methane liquefaction has enabled the problem of its delivery by sea from regions of production to those of use to be tackled; economically, it is more profitable to supply engeneering plants with argon and oxygen in a liquid state, to be then gasified at the place of use; oxygen and hydrogen as rocket fuel components are used only in a liquid state, since otherwise it is impossible to realize acceptable weight and size characteristics of on-board tanks. Using cryogenic products for cooling and cryostating various objects and devices gives still one more advantage of the liquid state, since at the same temperature the specific enthalpy (and internal energy) of the fluid is less than that of the vapor by the magnitude of the latent heat of vaporization.

As a result, the required temperature at the user's location is more easily attained and kept by a liquid phase.

Let us briefly consider upon the fields of application of each of the above-mentioned substances (CH_4, Ar, O_2, N_2, He) in liquid state.

1.2.1. Use of natural gas

Methane is the main component of natural gas widely used as fuel and raw material for the chemical industry.

Countries whose natural gas supply via pipeline is difficult or impossible because of the presence of long water obstacles, have developed, over the last decades, the technology of gas liquefaction and its transportation in a liquid state in special tankers. To do so, the relevant technological equipment was designed: powerful liquefying plants, tankers, equipped with thermally insulated vessels, large-scale storages, pipelines, pumps, gasifiers.

Methane-carrying tankers have cargo tanks 50÷100 m^3 or more in volume. The volume of the land tanks for storing liquid natural gas amounts to 130,000 m^3. Storing liquefied natural gas enables one to approach the problem of providing reserves, in order to supply regions during the periods of maximum use. At present, even in countries and regions to which natural gas is supplied via pipelines in gaseous state, a certain amount of the gas is liquefied to provide reserves.

Liquid methane is also used as an engine fuel. Its use allows the content of detrimental impurities in exhaust gases to be reduced and, at the same time, improvement in the economy of operating engines.

1.2.2. Use of oxygen

Oxygen is the most active oxidant, its reserves practically unlimited; it is non-toxic for man. This has dictated its wide use in a number of branches of industry: in nonferrous and ferrous metallurgy, chemical industry, rocket-space technology, mechanical engineering, medicine, etc.

Considerable amounts of liquid oxygen are used in rocket-space technology as a rocket fuel component.

In fact, use of liquid oxygen as the main fuel component in rocketspace technology has meant stipulated the development of large-scale liquid cryogenic systems which provide the assigned parameters to be accumulated, stored and obtained and liquid products to be delivered to the user. The advent of cryogenic systems providing rocket-space complexes and test beds with liquid oxygen has at the end of the fifties resulted in the head to design stationary and transport tanks, cryogenic pipelines of large length and high transmittivity, as well as other equipment (pumps, filters, armature, etc.); in considerable improvement of equipment specifications; in extensive studies of processes in liquid systems and in refinement of the technology connected with their service.

Oxygen filling systems are the largest-scale and most complex liquid cryogenic systems.

Thus, the space-rocket system "Saturn-5" was loaded with liquid oxygen to 1730 m^3 (2000 tons) from a tank 3230 m^3 in volume via pipeline 0.25 m dia and 580 m long.

Liquid oxygen is also used in on-board electrochemical generators and in self-support systems of vehicles and air-craft, in special MHD-generators. The spheres of its use in these systems are not wide but the systems themselves are extremely important for providing the functioning of apparatus and objects in which they are used.

In the majority of cases, gaseous oxygen is used in technological processes of metallurgical and chemical industry, in mechanical engineering, as well as in medicine.

Small-scale users are supplied with oxygen either in tanks or in pressurized recipients or in special reservoirs in liquid state, which is then to be gasified.

Supplying users with oxygen in liquid state simplifies its storage, and transport is therefore carried out on a larger scale.

This also refers to supplying users with other products of nitrogen and argon separation of the air. In the USA liquid oxygen amounting to 4 million tons and other air separation products (nitrogen and argon) are distributed between users through 5000 distribution stations. Liquid products are delivered to the distribution stations by tanker trucks and are drained into stationary reservoirs. Small- and medium-scale consumers are supplied with liquid products from stationary reservoirs of transport vessels and tanks. The system of supplying consumers with liquefied products of air separation has undergone considerable development for 15 ÷ 20 years in England; the storage volume has increased to 50,000 tons. In France, liquid oxygen is transported by tanker trucks to district gasification stations.

Other consumers (over 800) are supplied with liquid oxygen to be directly gasified at the location of the user. Also the Russia, supplying consumers with liquid oxygen, nitrogen and argon is under intensive development; the prime cost of oxygen delivered in liquid state with its subsequent gasification at the consumer's site is decreased $2.6 \div 6$ times as against its vessel delivery and $1.1 \div 2.4$ times as against its recipient delivery. Usually liquid cryogenic products are gasified through environmental heat but sometimes special heat carriers, hot water, vapor, are also used.

1.2.3. Use of argon

Argon is an inert gas and this property is used in mechanical engineering at metal welding and in the electric lamp industry. In liquid state, it is used exclusively when being delivered to consumers at a centralized factory supply.

1.2.4. Use of nitrogen

Because liquid nitrogen is non-toxic, inert and cheap, it is widely used as a cryogenic agent. Considerable amounts of liquid nitrogen are used in thermal vacuum chambers intended for simulating space conditions. In such chambers 1 to 100,000 m^3 or more, different aggregates, units and even assembled space systems are studied, tested and developed.

The space vacuum ($10^{-3} \div 10^{-8}$ Pa) is simulated by vacuum pumps, where, in large chambers, panels cooled by liquid or gaseous helium are used. Temperature conditions of outer space are simulated by mounting heat absorbing screens in the chamber, with a temperature close to that of liquid nitrogen.

The required screen temperature is maintained because of liquid nitrogen circulation. Cryogenic systems intended for providing space conditions in chambers are often characterized by high liquid nitroqen flow-rates. So, for a chamber 10,000 m^3 in volume, the circulating subcooled liquid nitrogen flow-rate through screens is approx. 10 $kg \cdot h^{-1}$ and the nitrogen volume is 2000 m^3. Liquid nitrogen is widely used for cold pressure tests of the oxygen, hydrogen and helium equipment.

The food industry represents the major consumer of liquid nitrogen. Fast cooling and freezing of foods by nitrogen spraying and their subsequent storage in nitrogen-enriched atmosphere preserve the taste quality and marketable state of products for considenable lengths of time.

Liquid nitrogen is also used in agriculture and medicine for storing bioproducts. Human blood frozen in liquid nitrogen may be preserved for months or even years. In medicine, liquid nitrogen is used in surgery for freezing when tumors, tonsils, and cataracts are being removed. In cryogenic systems, liquid nitrogen is widely used for cooling intermediate insulation screens of equipment, as well as for preliminary cooling of the large bulk of metal contained in of superconducting magnets, cables, transformers, etc. Liquid nitrogen is used for cooling varions resin, plastic and metal components, in order then to remove rolls, fins and so on. Liquid nitrogen cooling is used for embrittlement and subsequent crushing of scrap and worn resin items (particularly tyres). Gaseous nitrogen is used in a number of production processes in the chemical, petroleum chemical and metallurgical industries. In these cases, cold gasifiers are used to supply medium consumers with liquid nitrogen which is then gasified at the consumer's location. In particular, this technology of nitrogen supply is adopted to develop an inert atmosphere in hytrogen systems, in petroleum carriers, in some production processes of the chemical industry.

1.2.5. Use of hydrogen

Use of liquid hydrogen on an industrial scale began because of the development of rocket engines, for which cryogenic fluids, namely hydrogen and oxygen, are the fuel. The American space-rocket system "Saturn-5" took on-board 1275 m^3 ($\approx$ 90 tons) of liquid hydrogen which was then used at two upper stager of a carrier rocket. In the inter-orbital system "Shuttle", liquid hydrogen is also used as one of the fuel components.

Liquid hydrogen is used in bubble chambers for experiments on elementary particles. The large Soviet 70 GeV proton accelerator is equipped with a unique liquid hydrogen chamber approximately 11 m^3 in volume and about 3000 tons in mass.

Liquid hydrogen is also used in vehicle-borne electrochemical generators which transform the energy of a chemical reaction of adding oxygen to hydrogen immediately into electrical energy. Such electrochemical generators are used mainly in energy-supply systems for vehicles.

Considering the use of liquid hydrogen, it is necessary to consider briefly the promising uses of hydrogen in general.

At present (because of the scanty reserves of petroleum and natural gas), hydrogen is considered as a possible universal fuel for transport facilities. The attitude of specialists towards hydrogen is determined by the fact that it meets the requirements of an ideal fuel. In brief, such requirements are: fuel reserves must be inexhaustible; the cost must be low; a fuel must have a high specific energy-consumption, be convenient and safe in operation, not require essential modifications of existing type of engines, represent minimum danger to the environment during manufacturing, storage and burning. The advantages of hydrogen as a fuel are as follows. On combustion, it does not contaminate the environment, posesses high energy indices, its reserves are inexhaustible, its production is compatible with the production of the main kinds of energy: nuclear, thermal, electric; moreover, use of hydrogen requires minimal modification to existing systems. Drawbacks of hydrogen, that limit its use, are associated with its high cost, low density, the heed to design special equipment to liquefy and store it in liquid state.

Safety problems which appear when working with hydrogen are still complex but already a certain amount of experience has been gained in this respect. A general prejudice against the high danger associated with hydrogen use is slowly vanishing but may still slaken the pace of its future use. The greatest drawback of hydrogen as a fuel is its high cost: ways of reducing it are not yet clear.

The advent of a wider use of liquid hydrogen is connected, first of all, with its use as an aviation fuel. This specialist opinion is supported by a number of objective reasons, of which it is possible to distinguish those below. At present, aviation uses a comparatively small amount of petroleum but Its need is growing rapidly. The growth of consumption and relative high cost of traditional hydrocarbon fuels mean that simultaneous use of liquid hydrogen in aviation and hence extension of its production, are leading to some reduction in the cost of hydrogen.

Moreover, it should also be taken into account that liquid hydrogen use may lead to considerable improvements in flight characteristics of aircraft, e.g. increase their speed and decrease their weight.

1.2.6. Use of helium

Helium is widely used for cooling and cryostating a range of superconducting devices. Because of the relatively small heat evaporation and relatively high heat capacity of the vapor, as well as because of the narrow range of temperatures and pressures corresponding to the liquid state, along with liquid helium, use is also made of gaseous helium at parameters close to critical. Liquid helium for cooling and cryostasting superconducting devices is used in submerged-type cryostats and in circuits; gaseous helium is used in practice only in circuits. Recently, model and large-scale industrial systems of superconducting magnets, engines, generators, power lines, transformers, etc. which cooled and cryostated by liquid helium have been already designed and tested. Reseach is being carried out most intensively in the area of superconductivity for inducing powerful magnetic fields. Spheres of practical use of superpowerful magnetic fields include thermal nuclear Tokamak reactors, MHD-generators, elementary particle accelerators, transport.

Questions

1. Which cryoproducts are chemically inert and which are chemically active?
2. For which cryogenic substances does ice sink in the fluid?
3. What cryogenic substances are good heat carriers?
4. What are the magnetic properties of cryoproducts? What cryoagent is paramagnetic with high magnetic susceptibility?
5. What is the difference between ortho-hydrogen and parahydrogen? Of what does normal hydrogen consist and what is the ortho-para conversion?
6. Mention the main fields of application of different cryogenic products.
7. Characterize the cryogenic product from the viewpoint of ecological purity.
8. What are the safety rules when working with a cryogenic product?

2

HELIUM PROPERTIES: SUPERFLUIDITY

Natural helium is a substance consisting of two isotopes: He^4 (atomic mass of 4.0039) and He^3 (atomic mass 3.0017). Nowever, since the content of its lighter component is very low (atmospheric helium contains only 1.2×10^{-4} % He^3), the term "helium" is usually applied to the isotope He^4.

On the Earth, monatomic gas helium does not have wide use (5.24×10^{-4} % in total in atmospheric air) while worldwide the helium share is approximately 23% of the entire space mass.

Both helium isotopes are inert. Natural gas is the main source of He^4. Its content in the gas may amount to $1\div2$%. The content of the light isotope He^3 in He^4 usually constitutes $10^{-7} \div 10^{-6}$; therefore He^3 may be generated on the radioactive decay of tritium formed in nuclear reactors.

2.1. Helium properties

2.1.1. Thermodynamic properties of helium

Schemically, a **p-v-T** surface for He^4 is shown in Fig. 2.1. He^3 and He^4 have no specific properties in the gaseous states.

After hydrogen, gaseous helium has the next highest heat capacity, thermal conductivity and low density. Because of these properties and its neutrality, it is an excellent heat carrier.

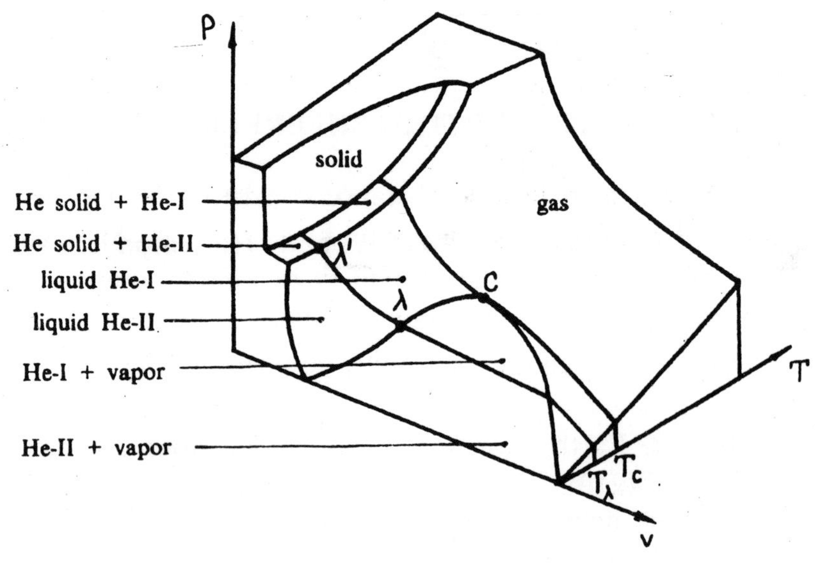

Fig.2.1. A p-v-T surface for He^4.

The **p-T** diagrams for He^4 and He^3 are plotted in Fig. 2.2. Point coordinates are given in Table 2.1 and Table 2.2. Dashed lines separate the regions with positive and negative volumetric thermal expansion coefficient β. In the state region where $\beta > 0$, liquid helium expands when heating. As follows from Fig. 2.2, helium has no triple point. A liquid phase exists up to absolute zero, and a pressure of more than 2.5 MPa must be applied to obtain a solid phase. While solidifying, the helium density changes slightly (its change is approx. 5%). Solid helium may also be obtained by compressing the gas. For example, the change of helium at T = 50 K to solid state needs a pressure of around 700 MPa. Three crystal modifications of solid He^3 and He^4 are known: with hexagonal (H), cubic volume-centered (CVC) and cubic faceentered (CFC) lattices.

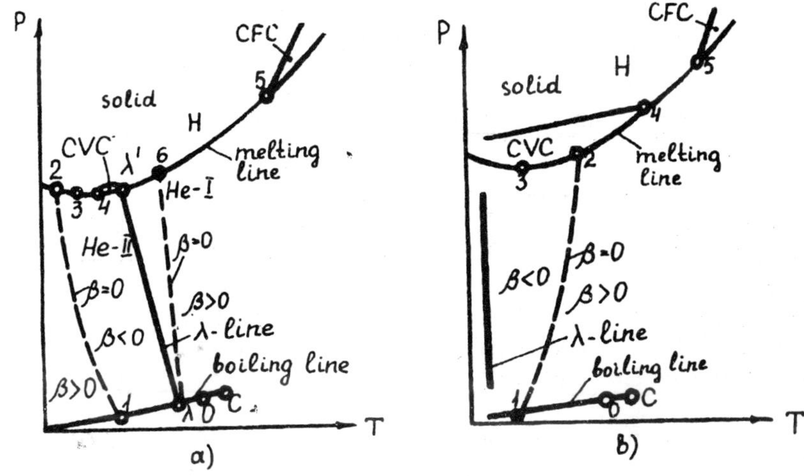

Fig. 2.2. Schematic p-T diagrams for He^4 (a) and He^3 (b).

Table 2.1. Parameters of characteristic points on the **p-T** diagram for He^4 (Fig. 2.2).

Characteristic points for He^4	T, K	p, MPa
C (critical point)	5.2014	0.2275
λ (lower λ-point)	2.172	0.005036
λ' (upper λ-point)	1.763	3.0134
0 (boiling point at p_0)	4.224	0.101325
1 (β = 0)	1.14	0.7093×10^{-3}
2 (β = 0)	0.59	2.5331
3 ($p = p_{min}$)	0.775	2.5291
4 (equilibrium He-II - CVC - H)	1.463	2.6273
5 (equilibrium He-I - CFC - H)	14.9	106.3912
6 (β = 0)	1.8	3.1309

The region of the He^4 liquid states is separated by the λ-line. To the right of the λ-line, liquid helium is in the normal (viscous) state typical of any liquid and denoted by He-I. To the left of the λ-line, liquid helium is in a special state, is called superfluid and denoted by He-II (Fig. 2.2).

Table 2.2. Parameters of characteristic points on the **p-T** diagram for He^4 (Fig. 2.2).

Characteristic points for He^3	T, K	p, MPa
C (critical)	3.324	0.1165
λ-line	0.003	-
0 (boiling point at p_0)	3.191	0.101325
1 ($\beta = 0$)	0.502	$0.2736\times10^{-}$
2 ($\beta = 0$)	1.26	4.7623
3 ($p = p_{min}$)	0.32	2.9303
4 (equilibrium liq. He^3- CVC - H)	3.138	13.7234
5 (equilibrium liq. He^3- CFC - H)	17.78	162.93

Transition from the region of He-I states to He-II states is called λ-transition and has been well studied for He^4, but for He^3, the λ-transition has only recently been revealed at $T \approx 0.003$ K.

The λ-transition occurs with no density jump (density curve of liquid in the vicinity of a λ-point has a sloping maximum), does not release or absorb the latent heat, and is accompanied by a sharply changing liquid helium heat capacity (Fig. 2.3 and Fig. 2.4). The helium transition across the λ-line is a second-order phase transition, whose distinctive feature is the discontinuity of the first temperature derivative of enthalpy. This means that in some temperature region, the specific helium heat capacity at constant pressure as a function of temperature is anomalous, i.e. $c_p \to \infty$.

The temperature T_λ, at which this anomalous behavior appears is called the λ-point, taking its name from the typical curve displaying liquid helium heat capacity as a function of temperature that resembles the Greek letter λ. Visual observations of the λ-transition yield the following typical flow pattern. While pumping-out helium vapors from the cryostat, the liquid boils within the entire volume, remaining to a considerable extent "light". Its temperature decreases gradually, and the boiling intensity decreases. This takes place because of the decreasing heat capacity of the liquid: the smaller the heat capacity, the less heat to be removed on cooling and the smaller the quantity of liquid that evaporates. After the temperature T_λ is attained the liquid boils up sharply, within the entire volume a large number of bubbles are formed, and the liquid-filled volume loses a large part of its transparency for a short time. Then the nucleation rate quickly decreases. After a short while, visible boiling ceases and the liquid becomes ideally quiet, and is He-II. If vapors continue to pump out, then the temperature may be reduced but not more by than 0.5 K.

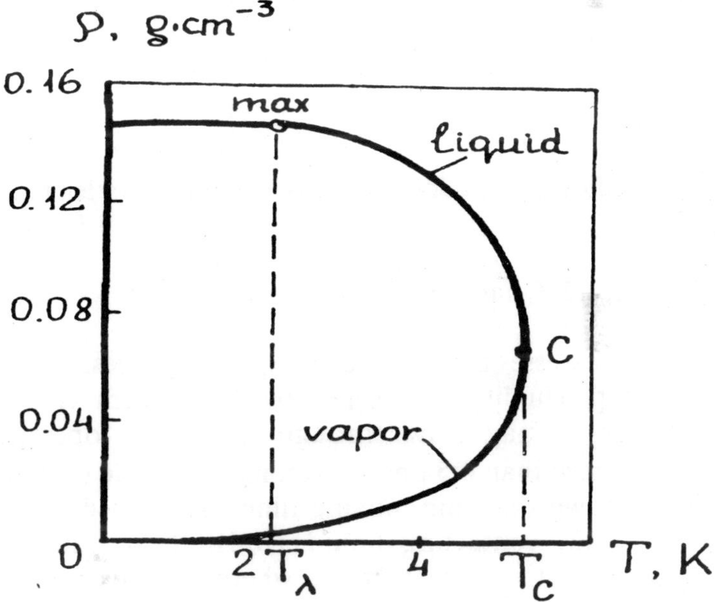

Fig. 2.3. He[4] density variation along the boundary curve for a two-phase fluid-vapor region.

This is attributed to the fact that the pressure becomes very small, and the amount evaporating liquid is insufficient to compensate further increase of the latent heat of evaporation and heat losses. When pumping out liquid He^3 vapor, a tem-perature of approximately 0.20 K can be achieved.

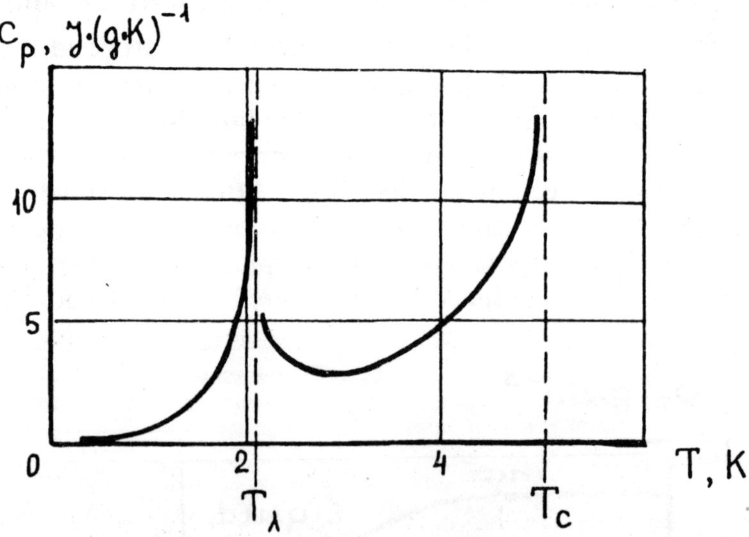

Fig. 2.4. Spesific heat capacity of liquid He^4 on the saturation line.

2.1.2. Quantum properties of helium

He-II possesses a number of unique properties, among which the superfluidity discovered by P. Kapitza is fundamental. He-II may flow through ultrasmall openings and slits not more than 0.5 μm in size, impenetrable even for a gas. It is known that thin moving films are formed on the surface of solids contacting liquid He-II. Such films are also available on the cryostat walls where He-II is present. The film moves up to a higher temperature. The film thickness is about 2×10^{-5} μm; e.g. the liquid helium level in a test tube (Fig. 2.5) is therefore set at one level with liquid He-II in the cryostat, as in the case when it was originally below and originally above the He-II level in the cryostat.

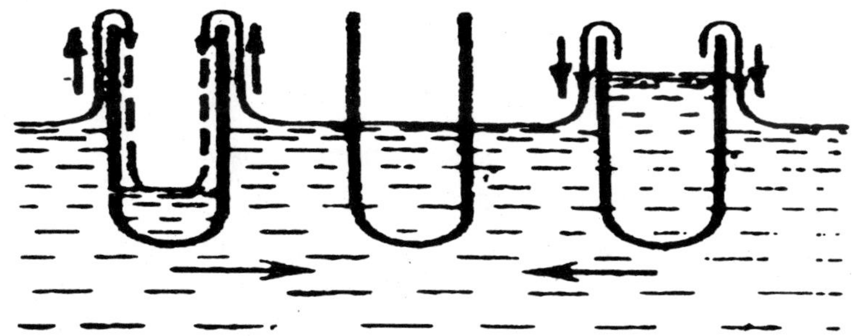

Fig. 2.5. Creeping He-II films.

An unusual behavior of heat transfer in He-II bound up with the indivisibility of heat and mass transfer processes results from the superfluidity.

In 1941, on the basis of P. Kapitza's experimental data, L. Landau developed the helium model and theory which enabled the superconductivity to be attributed to He-II superfluidity in the temperature range below the λ-point. He showed that in liquid He-II there exist two liquids with different properties: so-called normal and superfluid component, their ratio depending on temperature (Fig. 2.6).

The specific features of He-II are connected with the fact that He-II is a quantum fluid and its properties cannot be explained from the position of classical (using Newton's laws) mechanics but require a quantum mechanical description.

According to de Broglie's hypothesis, the wave properties of macroscopic bodies manifest themselves provided that $\lambda_D \geq l$, where λ_D is the de Broglie wavelength:

$$\lambda_D = \frac{h}{\sqrt{3mkT}};$$

l is the interatomic spacing:

$$l = \left(\frac{6m}{\pi\rho}\right)^{1/3};$$

h is Planck's constant; **k** is Boltzmann's constant; **m** is mass of the molecule; ρ is density of liquid.

Since for He-II at **T** = 2 K; $\rho \approx 146$ kg·m^{-3}; **m** = $=6.648\times10^{-27}$ kg; $l \approx 0.44$ nm and $\lambda_D \approx 0.89$ nm, i.e. $\lambda_D > l$ (at $T \to 0$ $\lambda_D \gg l$), it is clear that the He-II behavior must be determined by quantum effects. (As well as He-II, mixtures of He^3 and He^4, and He^3 are also treated as quantum fluid).

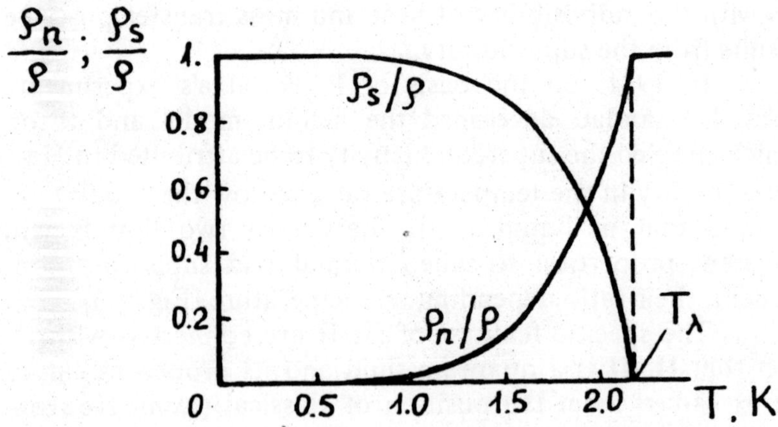

Fig. 2.6. Relative densities (mass fractions) of normal (n) and superfluid (s) components of liquid He-II as a function of temperature at saturated vapor pressure.

The modern quantum-mechanical description of He-II properties is based on the concept of a gas of quasi-particles (QP) developed by L. Landau in the years 1941-47. Liquid helium at He-II temperatures may be considered as a set of elementary excitations or a quantum system. Each of the elementary excitations may be described in the form of some quasi-particle moving inside the fluid. In this case, the number of excitations is assumed to be low, and their interaction energy to be much less than the intrinsic energy of quasi-particles. In other words, elementary thermal excitations may be considered to behave, in the known sense, as an ideal gas of QP.

At **T = 0**, the quantum system is in its lowest energy state. At a temperature different from zero, elementary thermal excitations may originate in a system, with its state interpreted as a set of arbitrarily interacting quasi-particles. The complete fluid energy in this case may be given as the sum of excitation energies of separate quasiparticles.

Bearing in mind that each of the quasi-particles moving inside the fluid is characterized by individual values of the energy ε and momentum **P**, the main characteristic of the QP gas is, as a whole, its energy spectrum $\varepsilon = \mathbf{f(P)}$, i.e. a set of energy levels. It should be emphasized that by the quasi-particle motion is understood not that of individual atoms but the result of excitation of the whole system.

The concept of elementary thermal excitations (concept of the QP gas) enables thermal properties of the macroscopic quantum system "liquid He" to be described. The energy spectrum of liquid helium (Fig.2.7) was postulated by L. Landau (1941-57) and, later on, around, 1959-61, was supported experimentally.

The He-II energy spectrum may be predetermined, proceeding from the averaged macroproperties, e.g. of the heat capacity of this fluid.

In particular, a knowledge of the heat capacity as a function of temperature opens the way to calculate the internal He-II energy; the latter allows a distribution function of the QP gas to be found, along with the energy spectrum $\varepsilon = \mathbf{f(P)}$.

The He-II energy spectrum was obtained in experiments with slow neutrons scattering. Measurements yielded the following spectrum parameters:

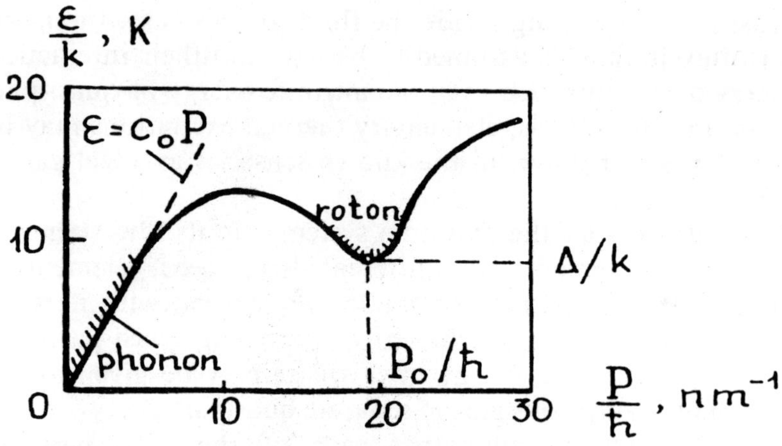

Fig. 2.7. An energy spectrum of liquid helium.

$$\Delta/k = 8.67 \pm 0.04 \text{ K};$$

$$P_0/\hbar = 19.36 \pm 0.5 \text{ nm}^{-1},$$

where $\mathbf{P_0}$ is the momentum at which the function ε has a minimum equal to Δ;

$$\hbar = h/(2\pi).$$

According to current thinking, there exist excitations of two kinds in liquid helium.

At $\mathbf{T \leq 0.6}$, the QP gas consists mainly of phonons (sound quants). A phonon section of the He energy spectrum is a straight line extending from the coordinate origin at a finite angle (Fig. 2.7). In this section, the energy as a function of momentum has the form

$$\varepsilon = c_0 P, \tag{2.1}$$

where $c_0 \approx 239 \text{ m}\cdot\text{s}^{-1}$ is the sonic velocity in helium.

At higher energies, the relation $\varepsilon = f(P)$ deviates from a straight line and passes successively through a maximum and a minimum. Thermal excitation with energies near the minimum of the curve $\varepsilon = f(P)$ are called rotons, thereby yielding a contribution, completing with photons, to all thermodinamic quantities.

For a spectrum roton section, the relation $\varepsilon = f(P)$ may be given as

$$\varepsilon = \Delta + (P - P_0)^2 / (2\, m_r)\,. \qquad (2.2)$$

The quantity m_r is usually called the effective roton mass.

Knowledge of the energy spectrum of liquid He-II enables its thermodinamic properties to be calculated with high accuracy, as well as account to be made for the mechanism of heat and mass transfer in liquid helium at low temperatures.

As regards this mechanism, it should first of all be noted that in thermal equilibrium when all directions P are equally probable, the mean momentum (per unit fluid volume) of the QP gasously, identically equal to zero, or the density of the mass flow of quasi-particles equals zero. In other words, at thermal equilibrium the QP gas involves no mass flow $j = \rho w = 0$.

Instead, with moving QP gas (in the presence of thermal excitations), the macroscopic momentum proves to be different from zero and proportional to a velocity w.

A proportionality factor ($kg{\cdot}m^{-3}$), is called the normal motion density ρ_n (or normal component density), and the momentum of unit volume of QP is, hence, the normal mass flow density j_n, $kg{\cdot}m^{-2}{\cdot}s^{-1}$). Then, the QP gas velocity may be called the normal velocity, and the relation may be written as

$$\vec{j}_n = \rho_n\, \vec{w}_n\,. \qquad (2.3)$$

Since in reality the vessel fluid is motioneless (i.e. visible macroscopic motion is absent in He-II), the total fluid momentum $\mathbf{j} = \mathbf{0}$, and hence mass $(\rho - \rho_n)$, must have another momentum equal to

$$\vec{j} - \vec{j}_n$$

and must have an opposite velocity.

Thus, with thermal excitations present in He-II, two internal macroscopic motions must exist simultaneously. The second internal motion is referred to as a superfluid motion. In this case:

$$\vec{j}_s = \vec{j} - \vec{j}_n$$

is the momentum of unit volume, or the superfluid mass flow density;

$$\rho_s = \rho - \rho_n$$

is the superfluid motion density (or superfluid component density);

$$\vec{w}_s = \vec{j}_s / \rho_s$$

is the superfluid velocity.

The density of the entire fluid is

$$\rho = \rho_n + \rho_s \,. \tag{2.4}$$

At motionless fluid temperature, only some part ρ_n / ρ of its mass is "subjected" to thermal excitations. Some amount of the mass

$$\frac{\rho_s}{\rho} = 1 - \frac{\rho_n}{\rho}$$

participating in the s-motion (superfluid motion) "has no" thermal excitations.

Unlike s-motion, n-motion (normal component) may be considered that of the viscous fluid.

It is obvious that at $T \to 0$ $\quad \rho_n \to 0$; $\rho_s \to \rho$;

at $T \to T_\lambda$ $\quad \rho_n \to \rho$; $\rho_s \to 0$.

The parameters ρ_n / ρ and ρ_s / ρ, in essence, characterize the He-II non-ordered and ordered states, respectively. The ratios ρ_n / ρ and ρ_s / ρ are sometimes therefore called non-orderliness (order-liness) parameters.

Thus, following current thinking, He-II may be conventionally considered as a set of two components: a superfluid component having zero viscosity, zero entropy and carrying no excitation quanta (i.e. as if it were at absolute zero) and a normal component that behaves like an ordinary liquid and possesses excitation quanta (phonons and rotons).

At $T = 0$ K, He-II as a whole consists of a superfluid component with density ρ_s, and at $T = T_\lambda$ only of a normal component with density ρ_n.

The mass flow density (total mass flow-rate) of a fluid at rest is equal to zero, i.e.

$$\rho_n \vec{w}_n + \rho_s \vec{w}_s = 0 . \tag{2.5}$$

In the case of moving He-II

$$\vec{j} = \vec{j}_n + \vec{j}_s \tag{2.6}$$

and its mean-mass velocity

$$\vec{w} = \vec{j} / \rho . \tag{2.7}$$

To a first approximation, ρ_n / ρ is approximated by $(T / T_\lambda)^7$, and ρ_s / ρ by the expression

$$\rho_s / \rho = 1 - (T / T_\lambda)^7 . \tag{2.8}$$

2.2. Manifestations of superfluidity

Consider, in more detail, a number of the effects in which helium superfluidity manifests itself:

1) mechanocaloric and thermomechanic effects at He-II flow in capillaries;

2) discs oscillating in He-II;

3) superfluid "wind tunnel";

4) He-II film flow.

2.2.1. Mechanocaloric and thermomechanic effects

Consider He-II flow in capillaries and small openings (slits). If two He-II vessels are connected by a capillary (or a slit) $1 \div 10$ μm in diameter and the fluid level difference (Fig. 2.8) is maintained, then helium will overflow from one vessel to the other even at $\Delta\mathbf{p} \rightarrow \mathbf{0}$, showing that the fluid viscosity is practically absent. The fluid flow-rate **G** through a capillary is obviously proportional to $\Delta\mathbf{p}/\mu$ i.e. where μ is the dinamic viscosity. Since $\Delta\mathbf{p} \rightarrow \mathbf{0}$ we have $\mu \rightarrow \mathbf{0}$, an indicator of the superfluidity property of He-II.

If for example excess pressure $\Delta\mathbf{p}$ is maintained, then because of fluid flow through the capillary, the temperature in this vessel will increase. This is attributed to the superfluid component with no excitation quanta moving in the capillary without friction (with zero viscosity).

The normal component cannot move in such a small-diameter capillary, its concentration in the right vessel elevates, thus increasing the temperature.

A temperature elevation will also be observed in the vessel from which He-II simply flows through a very small slit. The superfluid component flowing from the vessel does not carry entropy, and hence the remaining fluid possesses the same entropy but distributes it in a smaller mass, i.e. the fluid is being heated. An analogous explanation is valid for the cooling of the vessel into which the zero-entropy fluid flows.

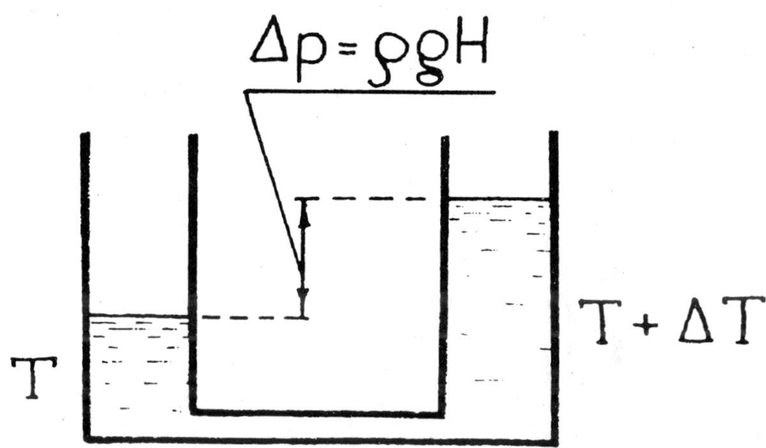

Fig. 2.8. Schematic of mechanocaloric and thermomechanic effect in He-II.

This phenomenon is referred to as a mechanocaloric effect: He-II heating (cooling) in a vessel when the the fluid flows through a capillary (or a small-diameter slit) from or into it.

Another classical example of the indivisibility of heat and mass transfer processes in He-II may be represented by the thermomechanic effect, where He-II starts moving when a temperature difference is applied. With heat supplied to one of the vessels, fluid (more accurately its superfluid component) starts overflowing through a capillary to this vessel, a level difference or a pressure difference $\Delta \mathbf{p}$ arises between the vessels. One more illustration of this effect is associated with so-called He-II spouting. One end of a wide tube (Fig. 2.9) filled with fine (e.g. emery) dust is immersed in a helium bath. Its other end is connected with a vertical capillary. On heating of the emery dust, the superfluid component flows into the fine convoluted channels formed by its fine particles, thus developing a level (pressure) difference, whose presence causes a He-II jet spouted from the capillary. It was found that for the equilibrium state or the conditions to reach it (quasi-equillibrium), the values of chemical potentials for one and the same fluid level of each of the vessels (Fig. 2.8) are identical, i.e.

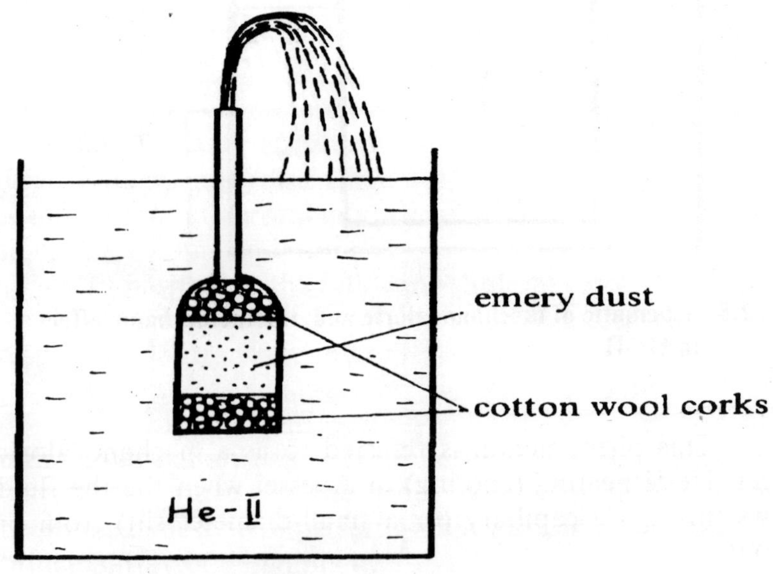

Fig. 2.9. He-II fountain.

$$\varphi(p_1, T_1) = \varphi(p_2, T_2), \tag{2.9}$$

or

$$\Delta\varphi = 0. \tag{2.10}$$

In other words, the driving force of the superfluid motion is a chemical potential difference, and the superfluid component moves down in a chemical potential (chemical potential in this case serves as a pressure analog).

From thermodynamics, it is known that a chemical potential differential is determined by:

$$\mathbf{d\varphi = v\ dp - s\ dT}\ , \qquad (2.11)$$

where **v** is specific volume; **s** is specific entropy.

Writing this formula in the from of increments, with regard to (2.10), we find that

$$\mathbf{\Delta p = \frac{s}{v}\ \Delta T = \rho\ s\ \Delta T}\ , \qquad (2.12)$$

where **Δp** is the thermomechanic pressure difference between He-II volumes with a temperature difference **ΔT**.

Expression (2.12) is referred to as London's formula. It shows that the onset of a temperature difference in He-II is always the cause of a pressure difference and, vice versa, and has good quantitative agreement with experiment. However, if the superfluid component velocity exceeds some critical value, relations (2.9) and (2.10) become invalid.

If a pressure difference is caused by a fluid level difference, i.e. $\mathbf{\Delta p = \rho\ g\ H}$, then

$$\mathbf{H = \frac{s\,\Delta T}{g}}\ . \qquad (2.13)$$

Calculations made using this formula show that if, for example, a temperature difference $\Delta T = 10^{-3}$ K is held between the vessels, then at $T = 1.9$ K a level difference constitutes approximatedy 7.5 cm. In other words, the superfluid component flow from the lower-temperature vessel to the hotter one will continue unless there appears an excess pressure $\mathbf{\Delta p} \approx 106$ Pa arises in the latter, (corresponding to a level difference **H** = 7.5 cm) which will halt further superfluid component flow.

2.2.2. Discs Oscillating in He-II

Experiments on torsional oscillation damping in liquid helium fully support the two fluid model of He-II.

It appears that if in He-II a circular thin aluminum foil disc is suspended via an elastic filament (Fig. 2.10) and small torsional oscillations around the filament axis are imparted to it, then the oscillations of this pendulum will be damped in nature. This clearly shows that He-II in this case behaves like an ordinary viscous (Newtonian) fluid. In an ideal non-viscous fluid, oscillation damping does not occur. From the oscillation damping rate, it is naturally possible to judge about the fluid viscosity.

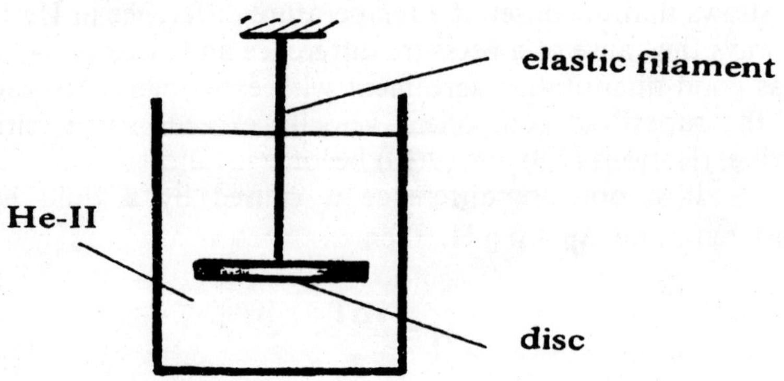

Fig. 2.10. Scheme illustrating experiment with an oscillating disc in He-II.

At the same time, using an oscillating pendulum of this type, E. Andronikashvili convincingly demonstrated the validity of the two component model when he, for the first time, measured ρ_n/ρ (hence, ρ_s/ρ) as a function of temperature and obtained the curves plotted in Fig. 2.6.

2.2.3. Superfluid "Wind Tunnel"

In this experiment, the ends of a cylindrical vessel placed into He-II (Fig. 2.11) are equipped with two partitions (inserts) made of a fine-dispersed emery dust. As already mentioned, these partitions have a large number of narrow, differently sized and irregularly shaped channels, and are impermeable for the viscous normal component.

Fig. 2.11. Illustration of an experiment with a fan in a superfluid "wind tunnel".

It is obvious that if some excess pressure **Δp** through the cylinder (called a superfluid "wind tunnel") is kept at one end of the cylinder, then the s-component alone will move. The normal component between the inserts will be practically motionless ($\mathbf{w_n} \approx 0$; $\mathbf{j_n} \approx 0$).

It is found that at small s-component velocities (less than some critical velocity equal to approx. 6 mm·s^{-1}), as expected, the s-component flow behaves like an ideal fluid (a fan set along the tube axis does not rotate). In other words, the s-motion at small velocities $\mathbf{w_s}$ interacts neither with the n-motion, tube walls, nor with the fan blades (viscous frictions is absent) and is, apparently, potential (irrotational), for which $\mathbf{w_s} = \mathbf{0}$. The slow s-motion thus induced may be preserved for an indefinitely long time.

Above the critical velocity, the flow behaves like an ordinary viscous fluid since the superfluid motion gives rise to a torsional moment that affects a fan. At such velocities, the superfluid motion is supposed to be turbulized; the s-component loses properties of an ideal fluid (interation starts with fan blades), i.e. superfluidity is lost.

2.2.4. He-II film flow

When formed on any solid surface in contact with the fluid, a thin helium film potentially moving along this surface is one of the most typical manifestation of He-II superfluidity.

The ability of fluids or saturated vapors to form an absorbed film on the surface of solids in contact with them is well known; however, because of the low value of the van der Waals forces in helium, a He-II film thickness noticeably exceeds that of adsorbed films of any other fluids. The mean helium film thickness is $(2.5 \div 3.5)\times 10^{-6}$ cm, i.e. about 100 atomic layers. To some (static) approximations, the He-II film thickness, δ, at a height $\mathbf{H}$ above the fluid level may be described by

$$\delta = \left(\frac{\alpha}{g\mathbf{H}}\right)^{1/3} \tag{2.14}$$

where α is a constant determined by the material and state of the solid surface. Such film thicknesses exclude flow of the He-II normal component but provide flow along the s-component film.

The presence of helium film on the He-II vessel walls may be the cause of considerable helium loss. This is well known to experimenters. In addition, the existence of a film increases the evaporation surface area of liquid helium and complicates the pump-down of its vapors.

Most characteristic examples of transfer of large fluid masses along a film are given by the experiments which have already become classical, where an empty vessel partially submerged into a He-II bath is filled with fluid up to the level of the bath of helium, or a He-II vessel filled and lifted above the helium bath is "spontaneously" emptied: liquid helium, moving in the form of film over the vessel walls, collects on the bottom surface and drops down, creating the impression that the vessel has sprung a leak.

Thorough experimental and theoretical research of moving He-II films has shown that as in the case of s-flow through narrow slits and capillaries, the chemical potential difference is the moving force that causes flow of the s-component along the film. This allows explanation of experimentally observed phenomena when a film moves from a site of lower temperature to one of higher temperature (a temperature gradient contributes to $\Delta\varphi$) or from a higher to a lower fluid level (a difference of gravitational potentials bet-ween the film ends is a driving force).

Exceeding some critical velocity of s-flow in a film, the flow potentiality is violated, energy dissipation appears, and superfluidity is lost.

Summarizing the above experimental findings, it may be concluded that s-motion appearing in He-II at small velocities w_s does not interact (exchange with momentum) with the walls, and n-motion, i.e. the s-component is in essence an ideal fluid. Different procedures to measure ρ_n, ρ_s allowing these parameters to be obtained as a function of temperature yield almost identical results and support Landau's two-fluid model.

However, in all cases considered, with a certain excess critical velocity, internal change in the s-flow occurs, resulting in viscous dissipative interaction with the walls and the n-component; the s-flow loses properties of an ideal fluid, and the superfluidity is lost.

The problem of determining the critical velocities in He-II was solved theoretically by L. Onsager (1949) and R. Feynman (1955). According to the Onsager-Feynman hypothesis, besides phonons and rotons, in He-II excitations connected with quantizing the s-flow circulation may exist.

Development of vortices in rotating He-II results in interacting **s** and **n** components, thereby causing critical phenomena.

Questions

1. What are the unique properties of helium that distinguish it from other substances?
2. What properties of helium make it convenient for technical use?
3. What is superfluidity? What are He-I and He-II?
4. What are the λ-transition, λ-point and λ-line?
5. Describe the concept of quasi-particles. What are phonons and rotons?
6. What is the mechanocaloric effect? How is this explained using the two-fluid model?
7. What is the thermomechanical effect? How is it explained?
8. Describe the damping of torsional oscillations in liquid He-II.
9. Describe an experiment with a superfluid "wind tunnel". What is the difference between processes at velocities below critical and above critical?

3

PROPERTIES OF SOLID BODIES AT LOW TEMPERATURES

Properties of solids, as well as of gases and fluids, depend strongnly on temperature, and change most at low temperatures. In practice, all physical properties of solids are determined by mutual location, of nature interaction, rate of motion of crystal lattice atoms and free electrons. If consideration is made of an ideal crystal, whose atom location displays a general rule at lattice points, then by using methods of classical statistical mechanics, expressions can be easily derived for internal energy, transfer coefficients, and other major characteristics.

However, for calculations at low temperatures such an approach yields incorrect results, and exact solution may be only by using quantum mechanics methods. Taking quantum-mechanical effects into account allows, in particular, analitical determination of heat capacity of solids. Along with calculation methods, wide use is made of expe-rimental means for determining properties of solids. Consider the physical essence of phenomena responsible for changes in these properties.

3.1. Heat capacity of solids

To determine heat capacity (of a solid $\mathbf{c_v} \approx \mathbf{c_p}$), it is necessary to know the magnitude of internal energy **U** ($J \cdot mol^{-1}$) and then to calculate molar heat capacity, $J \cdot mol^{-1} \cdot K^{-1}$:

$$\mathbf{C_v} = \frac{\partial \mathbf{U}}{\partial \mathbf{T}} \tag{3.1}$$

and specific heat capacity, $J \cdot kg^{-1} \cdot K^{-1}$:

$$c_v = \frac{C_v}{M} \qquad (3.2)$$

where M is molar mass ($kg \cdot mol^{-1}$).

3.1.1. Classical theory of heat capacity of solids

At high (room) temperatures, the internal energy may be determined by its uniform distribution in degrees of freedom. According to their distribution, energy $\varepsilon_1 = \frac{1}{2}kT$ accounts for one degree of freedom (k is Boltzmann's constant). For atom oscillations at the points of a crystal lattice, there are three degrees of freedom; if it is taken into account that a particle possesses both kinetic and potential energy, then the total energy amounts to $\varepsilon = 3kT$. For one mole of substance composed of N particles ($N = 6.022 \times 10^{23}$ is the Avogadro number) the total energy is

$$U = N\,3\,k\,T = 3\,R\,T, \qquad (3.3)$$

where $R = N\,k$ is a molar gas constant, and the heat capacity is

$$C_v = \frac{\partial U}{\partial T} = 3R.$$

At normal and higher temperatures, this result is well supported by experiment, and is consistent with the empirical law of Dulong and Petit.

3.1.2. Quantum-mechanical theories of heat capacity of solids

Another extreme case occurs at $T \to 0$; in this case, a corollary $C_v \to 0$ follows from the third law of thermodynamics. Thus, the heat capacity between room temperature and absolute zero must fall from 25 $J\cdot(mol\cdot K)^{-1}$ to zero. A change $C_v = f(T)$ is qualitatively well described by the Einstein theory of heat capacity of solids. Einstein was the first to use the quantum mechanics laws to determine the energy spectrum of atoms.

In this case, the mean energy ε_1 per degree of freedom of each oscillating atom is

$$\varepsilon_1 = \frac{h\nu}{\exp[h\nu/(kT)] - 1}, \tag{3.4}$$

where $h \approx 6.62\times10^{-34}$ J·s is the Planck constant; ν is the frequency of oscillations, s^{-1}.

The total energy of one mole

$$U = 3N\varepsilon_1;$$

then based on the relation

$$C_v = \frac{\partial U}{\partial T},$$

an expression may be written for heat capacity

$$C_v = 3R\left(\frac{\Theta_E}{T}\right)^2 \frac{\exp(\Theta_E/T)}{[\exp(\Theta_E/T) - 1]}, \tag{3.5}$$

where $\Theta_E = h\nu/k$ is the characteristic temperature.

Formula (3.5) agrees well with experiment at moderately low temperatures. A quantitative inaccuracy of Einstein's formula at low T results from the fact that the oscillation frequency ν was assumed constant.

Debye theory of heat capacity of solids. Debye developed a more exact theory by taking account of the difference of oscillations ν and discontinuity of their spectra. In this case, the possible number of all modes of oscillations is equal to **3N**. Having set a limit of a maximum frequency ν_m on the spectrum, Debye determined the energy of an oscillating system like this and then its heat capacity, J·(mol·K)$^{-1}$:

$$C_v = 3R\left(\frac{T}{\Theta_D}\right)^3 f_D\left(\frac{T}{\Theta_D}\right), \tag{3.6}$$

where the Debye function is

$$f_D\left(\frac{T}{\Theta_D}\right) = 3\int_0^{\Theta_D/T} \frac{x^4 \exp(x)}{[\exp(x)-1]^2} dx.$$

Here $x = h\nu/(kT)$.

In this theory, an important quantity Θ_D is the Debye temperature:

$$\Theta_D = \frac{h\nu_m}{k}. \tag{3.7}$$

The quantity Θ_D is a parameter for each substance, and has the physical meaning of a temperature boundary, below which the heat capacity starts changing essentially because of varying **T**. For high temperatures ($T >> \Theta_D$) from expression (3.6) we obtain, in the limiting case, the classical result: $C_v = 3R$. At very low temperatures ($T << \Theta_D$) starting practically with $T < 0.1\,\Theta_D$ we have another limiting case; the heat capacity, J·(mol·K)$^{-1}$ is

$$C_v = 234R\left(\frac{T}{\Theta_D}\right)^3, \tag{3.8}$$

i.e. heat capacity change is proportional to temperature raised to the third power.

Debye's theory agrees well with experimental data and may be the basis for calculations of the heat capacity of many materials. Values of the function **f(T/ΘD)** are known, and calculations using formula (3.6) give diagrams or tables. In Fig. 3.1 the heat capacity $\mathbf{C_v}$ is plotted, expressed in units of gas constant **R** as a function of the dimensioneless quantity T / Θ_D. Values of the Debye temperature Θ_D for some substances are listed in Table 3.1.

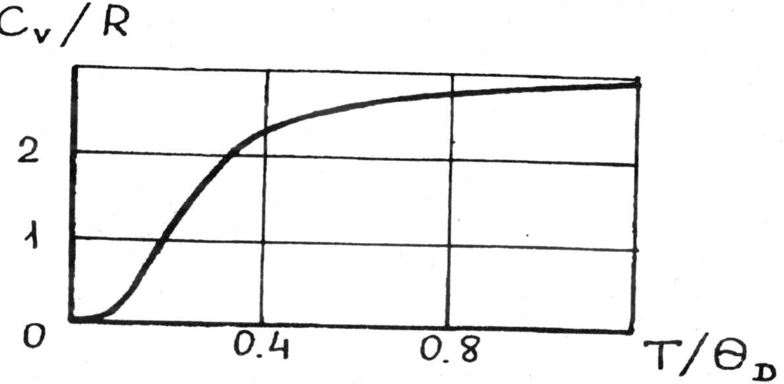

Fig. 3.1. Dimensionless heat capacity of solid bodies as a function of T/Θ_D.

Table 3.1. Values of Θ_D for some substances.

Substance	Al	Fe	Cu	Zn	Sn	Ag	Pb	C graphite	C diamond
Θ_D, K	420	467	343	305	190	225	95	391	2000

These data point to a wide range of Θ_D. So, for lead Θ_D has a small value, and hence the ratio T / Θ_D is large and the heat capacity starts decreasing at lower temperatures: the inverse relationship is typical of diamond. Nowever, relation (3.6) is not always strictly satisfied; at low temperatures **T**, additional contributions to the heat capacity and its anomalous changes are possible. So, at a temperature of several K the energy bound to the motion of free electrons in metals becomes noticeable. The electron component of the heat capacity C_{ve} corresponding to this energy may be calculated by using the concept of a degenerated electron gas obeying Fermi-Dirac statistics.

The value of this component is proportional to the first degree of temperature:

$$C_{ve} = R\frac{\pi^2}{2}\frac{T}{\Theta_0}, \tag{3.9}$$

where C_{ve} is measured in J·(mol·K)$^{-1}$; Θ_0 is the characteristic temperature.

At sufficiently low temperatures, the value of electron heat capacity C_{ve} may exceed the heat capacity C_{vl} because of lattice atom oscillations, since the latter is proportional to temperature raised to the cubic power in agreement with (3.8).

Fig. 3.2 gives a comparison of the components of aluminum specific heat capacity. At $T < 4$ K, c_{ve} is seen to exceed c_{vl}.

In cryogenics, paramagnetic substances used in adiabatic demagnetization systems are significant. At low temperatures, themagnetic dipole interaction energy ε_m becomes comparable with the atom oscillation energy, and the magnitude ε_m specifies the main contribution to heat capacity. The corresponding magnetic component of the heat capacity C_m changes in a complex manner, thereby increasing to a maximum at a characteristic temperature $\Theta_m = \varepsilon_m / k$ and then decreasing to zero at $T \to 0$. At $T > \Theta_m$ the magnetic heat capacity varies according to the square law.

$$C_m = \frac{R}{4}\left(\frac{\Theta_m}{T}\right)^2 . \qquad (3.10)$$

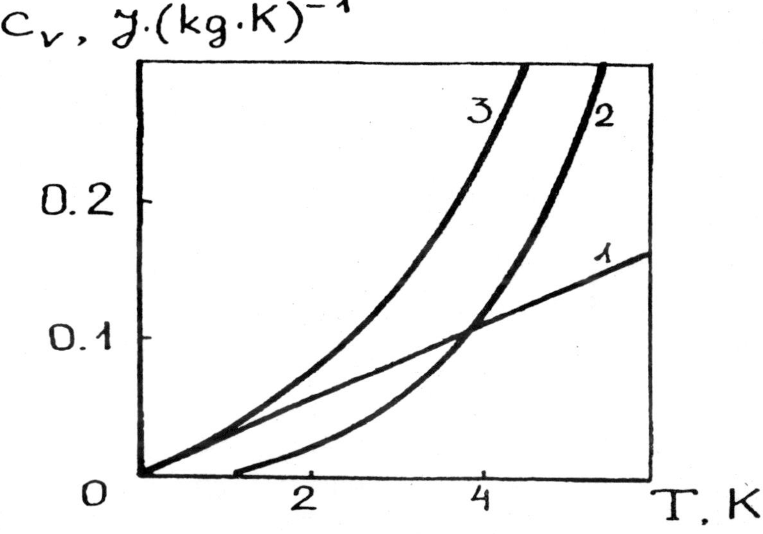

Fig. 3.2. Comparison of electron (1), lattice (2) and total (3) heat capacities at low temperatures.

3.2. Thermal conductivity of solids

The thermal conductivity λ, $W\cdot(m\cdot K)^{-1}$ characterizes the mechanism of energy transfer which in a solid is specified by the free electron motion and the crystal lattice oscillation. The redominance of a certain mode of transfer depends on the type of material and its properties. In conductors, heat is transferred by electrons, whilst with intermediate properties mixed transfer takes place. Thus, in the general case

$$\lambda = \lambda_e + \lambda_l ,$$

where λ_e and λ_l are the electron and lattice (or phonon) thermal conductivity, respectively.

The phenomenon of electron thermal conductivity λ_e is interpreted using the model for a free electron gas where electrons participate in random thermal motion like ideal gas molecules. The electron gas may also be described by a free electron path length $\mathbf{l_e}$ and a mean velocity of electron thermal motion $\mathbf{w_e}$ ($m \cdot s^{-1}$).

To describe lattice heat conduction (or phonon heat conduction) due to the propagation of cristal lattice oscillations, it is convenient to adopt the phonon gas concept. The propagation of thermal waves bound to the energy of elastic lattice oscillations may be considered as the motion of quasiparticles called phonons. The phonon energy is quantum in nature, determined by an oscillation frequency ν and is constant for such phonons. The phonon velocity must be equal to that of propagating elastic oscillations in a solid, i.e. sonic velocity. Unlike atoms and electrons, the number of phonons is not constant and increases with rising temperature. Phonon motion involves phonon collisions and scattering of their energy; hence, there exists the concept of free phonon path length.

Thus, the motion of electron and phonon gas particles is responsible for heat conduction in a solid. This fact allows use of the heat conduction equation obtained from molecular-kinetic theory of an ideal gas applied to heat conduction in a solid. With this in mind, we consider a solid as a volume filled with electrons and phonons.

Such an approach does not yield an exact quantitative result; however, it qualitatively explains the thermal conductivity variation at low temperatures. Using the formula for heat conduction in a gas

$$\lambda = \frac{1}{3}\, \mathbf{l}\, \mathbf{w}\, \rho\, \mathbf{c_v}\,, \tag{3.11}$$

a change of each factor present may be determined as a function of **T**.

Electron thermal conductivity λ_e is typical of metals where electrons almost completely define a heat transfer process and $\lambda_l \rightarrow 0$. The behavior of separate multipliers present in this formula displays the relation plotted in Fig. 3.3.

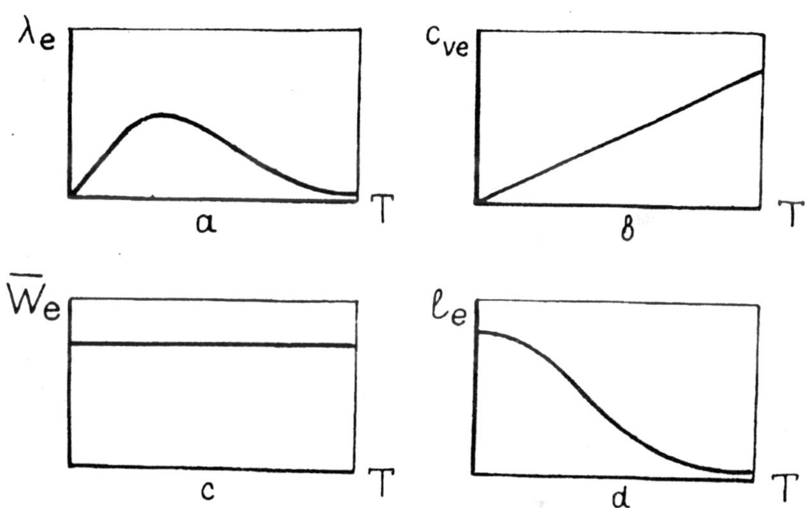

Fig. 3.3. Electron thermal conductivity versus temperature: a, resulting curve; b, electron heat capacity; c, electron velocity; d, free path length.

Heat capacity per unit volume of an electron gas (ρ c_{ve}), J·(m³·K)⁻¹ determined through formula (3.9) linearly decreases with decreasing temperature (curve **b**). An electron velocity near the Fermi level in practice does not depend on temperature: w_e = **const** (line **c**). The free electron path l_e is maximum at low temperature but decreases with increasing **T** (curve **d**), as in this case the number of phonons increases and the number of collisions with electrons grows.

Interaction of opposing factors gives rise to a maximum in the resultant curve **a**. At $T \to 0$, the heat capacity $c_{ve} \to 0$, and thermal conductivity also takes on a zero value. The presence of lattice impurities and defects, in turn, results both in additional scattering of electron energy and in a decrease of free path length; as a result, thermal conductivity also decreases. Thus, the degree of purity of a material and the way it is finished affect thermal conductivity which has higher values for pure and non-deformed metals.

Lattice thermal conductivity λ_l. Phonons play a determining role in the heat transfer of non-metals. Analysis of the components of a general formula for a phonon gas enables one to reach the following conclusions. The heat capacity change is determined by Debye's formula (3.6) (see Fig. 3.1); at $T = 0$ thermal conductivity will also be obviously equal to zero. Phonon velocity w equals the sonic value, varying slightly with changing temperature. The free path length is maximum at $T = 0$ and decreases sharply with increasing T, resulting from the increasing number of phonons and their collisional frequency. The resultant curve for phonon thermal conductivity, as well as for electrons, has a maximum whose value is different depending on the characteristics of a material. The lattice defects decrease the free path and value of λ_l. Values of λ_l are especially low for disordered structure materials (glass, plastics); in this case, the maximum moves up in temperature.

Values of λ_l and λ_e of materials possessing intermediate properties have the same order, and the resultant thermal conductivity curve keeps the same shape with a characteristic maximum.

Fig. 3.4 plots the experimental thermal conductivity of some pure metals; maxima are clearly seen and are found in the low temperature region. For non-metals, the maxima move up in temperature (Fig. 3.5).

3.3. Thermal expansion of solids

The thermomechanical effect of thermal expansion is of great significance for materials at cryogenic temperatures. A body's thermal expansion is connected with the change in

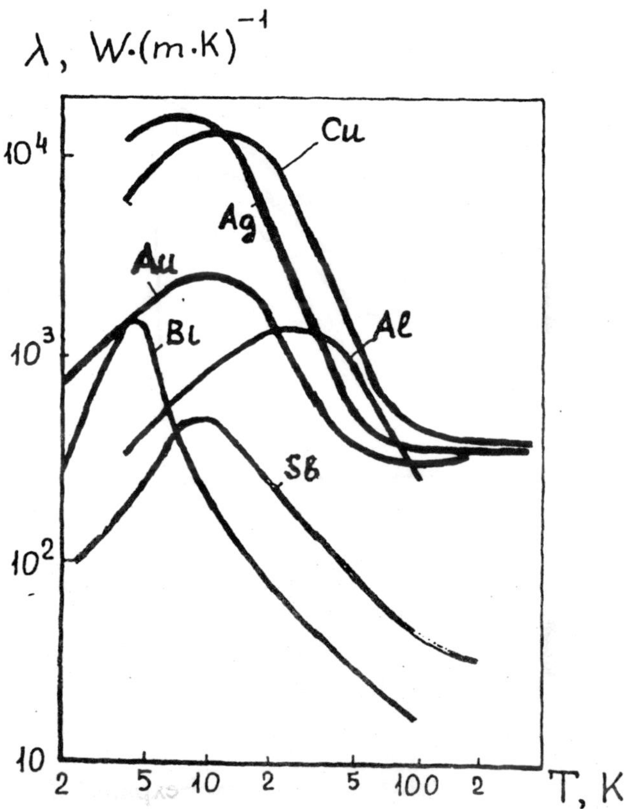

Fig. 3.4. Experimental thermal conductivity of some pure metals as a function of temperature.

position of atoms at different energy levels. Thus, near absolute zero, the energy is minimum and the atom position is equilibrium; the distance between atoms is invariable and equal to $\mathbf{r_0}$.

With increase in temperature, the vibrational energy of atoms grows and the distances between atome specifying their equilibrium position increase. As a result, the body expands.

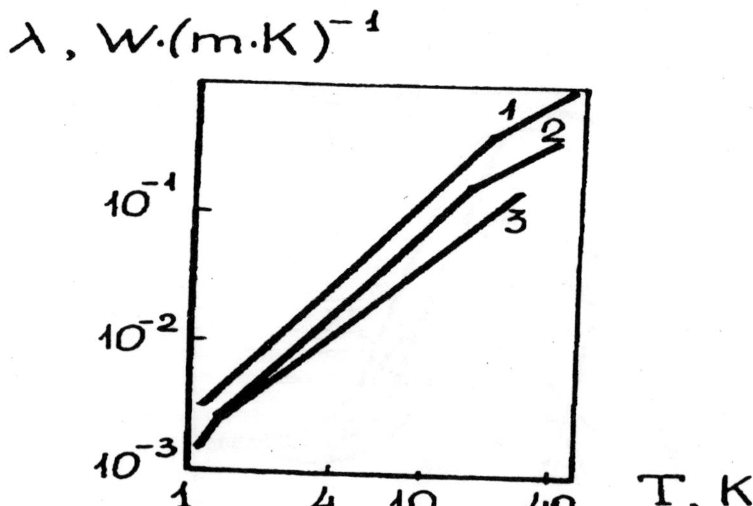

Fig. 3.5. Thermal conductivity of widely used non-metals as a function of temperature.
1, polyethylene with $\rho = 971$ kg·m^{-3}; 2, polyethylene with $\rho = 956$ kg·m^{-3}; 3, nylon.

The quantitative value of the thermal expansion effect is set by using the temperature coefficient of volumetric expansion,

$$\beta = \frac{1}{v}\left(\frac{\partial v}{\partial T}\right)_p . \qquad (3.12)$$

For solids, also, wide use is made of the concept of the temperature coefficient of linear expansion:

$$\alpha_T = \frac{1}{L}\left(\frac{\partial L}{\partial T}\right)_\sigma , \tag{3.13}$$

where L is the linear dimension and σ is the internal stress in a material.

For isotropic substances, $\alpha_T = \beta/3$. The behavior of coefficients α_T and β as a function of temperature is determined as follows.

From (2.11), it follows that the partial derivatives of a chemical potential $\varphi(p,T)$ are determined by the formulas:

$$\left(\frac{\partial \varphi}{\partial p}\right)_T = v ; \qquad \left(\frac{\partial \varphi}{\partial T}\right)_p = -s . \tag{3.14}$$

With the derivative

$$\frac{\partial^2 \varphi}{\partial p \, \partial T}$$

from equalities (3.14), the Maxwell relation

$$\left(\frac{\partial v}{\partial T}\right)_p = -\left(\frac{\partial s}{\partial p}\right)_T \tag{3.15}$$

is obtained, and is convenient for analyzing the coefficients α_T and β in the low temperature region.

At $T \to 0$, according to the third law of thermodynamics

$$(\partial s/\partial p)_T \to 0 .$$

It is obvious that in this case,

$$\beta = \frac{1}{v}\left(\frac{\partial v}{\partial T}\right)_p \to 0$$

and $\alpha_T \to 0$. With increasing temperature, α_T increases. An approximate value of the coefficient β may be found by the empirical Gruneisen's formula

$$\beta = \gamma_g \kappa_T \rho c_v , \tag{3.16}$$

where γ_g is the Gruneisen constant (for many materials its value is close to two) and κ_T is the temperature coefficient of isothermal compressibility.

Quantities γ_g and κ_T depend slightly on temperature, and hence a change in coefficient β is mainly attributed to the heat capacity c_v behavior. This enables one to find approximate values of the coefficient β to be found through the relation

$$\frac{\beta(T)}{\beta(T_0)} = \frac{c_v(T)}{c_v(T_0)} , \tag{3.17}$$

where $T_0 = 293$ K is room temperature and T is the temperature at which a desired value of β is to be found.

Rapidly decreasing the temperature coefficient of expansion with decreasing temperature results in the fact that at a temperature below 50 ÷ 80 K, deformation becomes barely observable.

Frequently, practical calculations use the concept of a mean temperature coefficient of linear expansion over a temperature range ΔT; then the linear dimension change due to thermal deformation is

$$L(T) = L_0 (1 + \alpha_T \Delta T) , \tag{3.18}$$

where L_0 is the initial dimension and α_T is the mean value over a temperature range ΔT.

In Fig. 3.6 is plotted the relative thermal deformation $\Delta L / L_0$ versus temperature; the curve behavior shows that at temperatures below 50 K, dimensions practically cease to change.

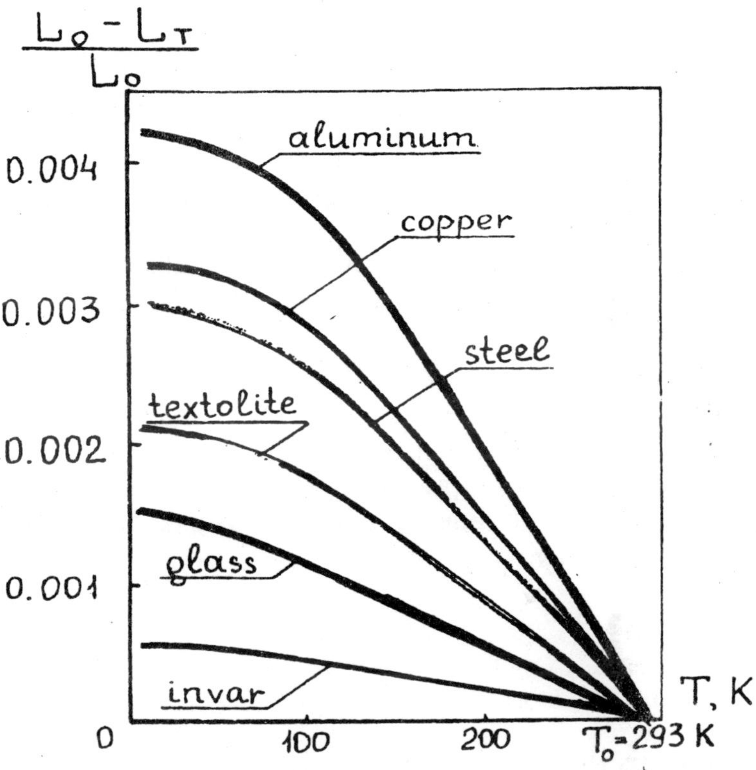

Fig. 3.6. Curves for relative thermal deformation of various materials at low temperatures.

3.4. Thermal radiation of solids

Thermal radiation is a process of energy transfer by electromagnetic waves. Thermal radiation is one of the fundamental mechanisms of heat transfer in cryogenic thermal insulation and in vacuum cryogenic equipment. At temperatures below 300 K, thermal radiation proceeds in the infrared region of a spectrum at wavelengths $l_w = 10\ \mu m$ and over. The emissivity of substances is characterized by a thermal radiation coefficient

$$\varepsilon = \frac{E}{E_s}, \tag{3.19}$$

where **E** is the radiation flux density of a given body; $E_s = \sigma T^4$ is the radiation flux density of an absolutely black body at the same temperature where σ is the Stefan-Boltzmann constant.

Following Kirchhoff's radiation law, the emissivity ε is equal to the absorptivity a of a body at given wavelength and temperature. Studying the optical properties of metals by the classical electron gas theory allows an approximate expression to be obtained for absorptivity a and thermal radiation coefficient ε :

$$\varepsilon = a \approx 0.365\sqrt{\frac{\rho_e}{l_w}}, \tag{3.20}$$

where ρ_e is the specific electric resistance, $\Omega \cdot m$; l_w is wavelength, m.

Table 3.2 gives emissivity ε values for some materials at various temperatures.

The mechanism for interaction between radiation and a metal surface indicates that radiation excites forced oscillations of free electrons, thereby giving rise to a reflected wave. The damping of electromagnetic waves in a metal surface layer, accompanied by reflection of some part of the energy flux, specifies the absorptivity of a body. Relation (3.20) shows that the absorptivity and emissivity of a substance depend strongly on its electrical conductivity $1/\rho_e$. Low-absorptivity materials must have a small value of ρ_e, i.e. they must be good current conductors. As ρ_e decreases proportionally to temperature, the emissivity decreases with lowering **T**.

Table 3.2. The emissivity of solids.

Material	Emissivity ε at temperature, K		
	300	76	4.2
Aluminum, annealed (electrically polished)	0.03	0.018	0.011
Copper (mechanical polishing)	0.03	0.019	0.015
Copper (electrolytic polishing)	-	0.015	0.0062
Silver	0.022	0.01	-
Steel, corrosion-resistant	0.1	0.048	-
Carbon steel	0.6	-	-
Glass	0.94	-	-

At low temperatures, formula (3.20) does not yield a quantitatively correct result because of the appearance of the additional surface effect (skin effect) connected with increasing free electron path length. Real values of emissivity prove to be much higher and the temperature effect breaks down.

The material surface state is another important fact affecting ε. Rupture of the ordered crystal structure in a surface layer reduces electrical conductivity and ε increases. Values of ε are minimum in relieving surface layer stresses (annealing, etching); mechanical polishing is the cause of the ε increase. The surface emissivity is also of great importance.

3.5. Electric properties of solids

Electric properties of solids depend strongly on temperature. Electric charge transfer in conductors results from simultaneously acting free electrons in an electric field. During their movement, conduction electrons moving under the action of an electric field encounter obstacles that cause the scattering of their energy, i.e. a decrease in electrical conductivity σ_e and an increase in specific electrical resistance $\rho_e = 1/\sigma_e$.

Such obstacles may be thermal lattice oscillations (phonons), or else lattice defects due to impurities and crystal lattice ruptures. As a result of collisions with phonons and lattice defects, the electron velocity drops and then grows when an electric field is applied. It is obvious that such a mechanism of charge transfer depends strongly on the mean electrons velocity $\mathbf{w}_e$ and mean path length $\mathbf{l}_e$.

The classical theory of electrical conduction dealing with ideal electron gas flow and gas interaction with phonons defines the electrical conductivity as:

$$\sigma_e = \frac{1}{\rho_e} = \frac{ne^2 l_e}{2mw_e}, \tag{3.21}$$

where $\mathbf{n}$ is the number of free electrons per 1 m³ of material; $\mathbf{e}$ is the elementary charge; $\mathbf{m}$ is the electron mass.

It is evident that electrical conductivity as a function of temperature is determined by varying the free path length l_e (Fig. 3.3d).

Heat and current transfer in conductors is of the same nature and is effected by the electron gas; therefore, thermal conductivity λ_e and electrical conductivity σ are connected by a simple relation.

For an ideal gas, the unit volume the heat capacity may be determined as:

$$c_v \rho = \tfrac{3}{2}\, n\, k . \qquad (3.22)$$

Substituting this expression into (3.11) for thermal conductivity and bearing in mind that

$$\frac{m w_e^2}{2} = \tfrac{3}{2}\, k\, T ,$$

from expression (3.21) we have

$$\frac{\lambda_e}{\sigma_e} = 3\left(\frac{k}{e}\right)^2 T . \qquad (3.23)$$

This formula is the experimentally known Wideman-Frants rule. It is well satisfied at temperatures above 100 K; however, at lower temperatures, deviations are observed since the free path length is different for λ_e and σ_e.

The more exact electrical conduction theory is based on the quantum mechanics laws used to describe properties of a collective system, or an electron gas interacting with a crystal lattice. This theory uses Fermi-Dirac statistics obeyed by the electron gas. A more exact result in calculating the electrical resistance is given by Bloch-Gruneisen formula, which has as a corollary the following relation

$$\frac{\rho_e(T)}{\rho_e(\Theta_D)} = 1.056\left(\frac{T}{\Theta_D}\right)^5 F\left(\frac{\Theta_D}{T}\right), \qquad (3.24)$$

where $\rho_e(T)$ and $\rho_e(\Theta_D)$ are the specific electrical resistance of a conductor at arbitrary temperature T and at the Debye temperature given by formula (3.7); $F(\Theta_D / T)$ is a function whose values are in tabular form.

The form of function (3.24) is shown in Fig. 3.7. At low temperatures $(T < 0.1\, \Theta_D)$, the electrical resistance decreases rapidly in proportion to T^5, and at high temperatures ($T > 0.5\ \Theta_D$) varies in direct proportion to T.

Determining total electrical resistance ρ_t, account must also be taken of another of its components ρ_0, which

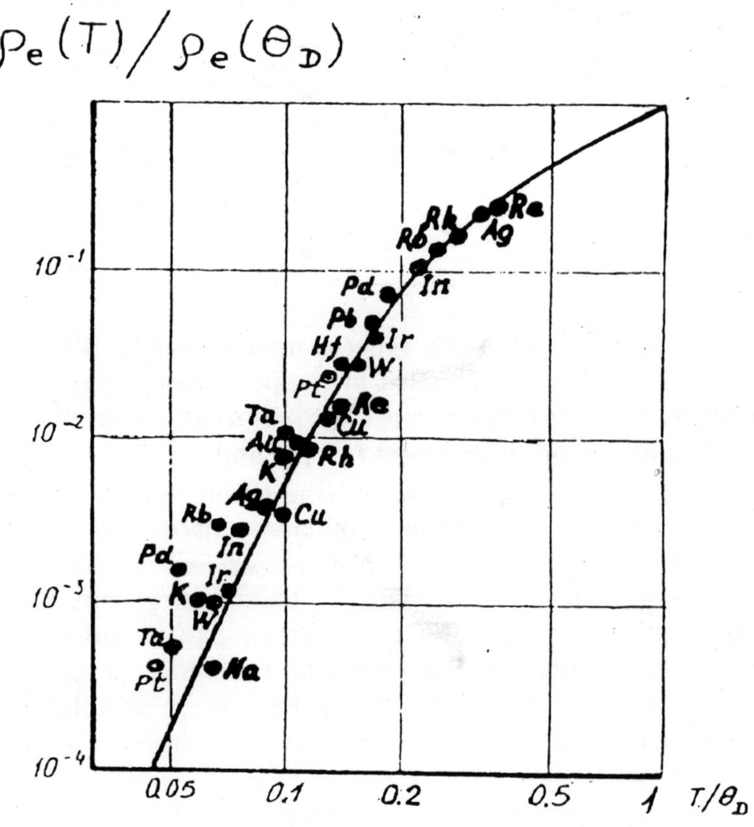

Fig. 3.7. Reduced electrical resistance versus relative temperature.

is due to energy scattering on impurities and lattice defects. This component does not depend on temperature and yields a residual resistance, up to $T = 0$.

For good conductors $\rho_0 << \rho_e$; the magnitude ρ_0 starts manifesting itself at low temperatures (80 K and below) and is determining at temperatures of the order of 10 K. The influence of ρ_0 grows with increasing impurities and defects.

According to Mathiessen's rule, the total electrical resistance is

$$\rho_t = \rho_e + \rho_0 \, . \tag{3.25}$$

Fig. 3.8 show a plot of specific resistance against temperature.

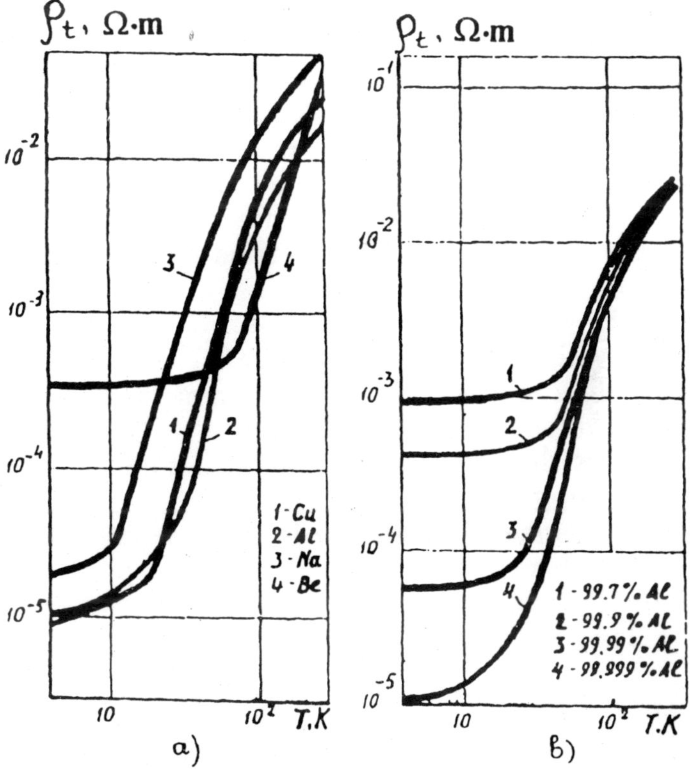

Fig. 3.8. Specific electrical resistance versus temperature: a, various metals; b, aluminum with varous degrees of purity degree.

Horizontal fragments of the curves correspond to residual resistance ρ_0. The above applies to conductors of sufficiently high purity. Electrical conductivity of melts is mainly determined by impurity concentration, i.e. by the quantity ρ_0. The electrical resistance of a melt depends on temperature to a lesser degree than for pure metals.

Questions

1. What is the significance of the Dulong and Petit law? What is the theoretical deduction of this law?
2. What are the main ideas of the Einstein and Debye theories of the heat capacity of solids?
3. Describe the phonon, electron and magnetic components of heat capacity.
4. What is the main difference between the physical mechanism of heat conduction of solid dielectrics and that for metals?
5. How and why does the electron thermal conductivity vary with decreasing temperature?
6. How and why does the phonon thermal conductivity vary with decreasing temperature?
7. Describe the variation in temperature coefficient of linear expansion at low temperatures.
8. Describe the variation in thermal radiation coefficient at low temperature.
9. How and why does the electrical conductivity of solids at low temperatures vary?

4

CLASSIFICATION OF CRYOGENIC SYSTEMS

From the viewpoint of thermodynamics, two types of processes are possible which result in cooling a substance:

1) processes occurring with heat supply to another medium (external cooling);

2) processe occurring without heat removal, e.g. throttling or expansion (internal cooling).

To bring about external cooling, an external system is needed, whose temperature is below that of the object to be cooled; as a result, the required heat transfer process takes place. Internal cooling is connected with decreasing the value of some intensive state parameter such as pressure, chemical potential, electrical field strength or magnetic field intensity. In any low-temperature system, the lowest temperature can only be attained by internal cooling.

Thus, at temperatures below ambient internal cooling processes are of most significance, thus providing the necessary conditions for external cooling processes to proceed.

4.1. General principle of internal cooling

Consider the general principle of internal cooling of an arbitrary system. As is known, the thermodynamic system entropy $\mathbf{s}$ depends on a system temperature $\mathbf{T}$, as well as on a number of other parameters, namely pressure, intensity of magnetic and electric fields, etc. In the general case,

$$\mathbf{s} = \mathbf{s}\,(\mathbf{T}, \mathbf{x}_1, \mathbf{x}_2, \ldots, \mathbf{x}_n)\,,$$

where the $\mathbf{x}_i$'s are the independent variables characterizing the possible type of interaction (mechanical, magnetic, electric, etc.) between the considered system and those external to it.

Choose one of these parameters x_i by fixing all the remainder. The slope of the curves x_i = **const** in **T-s** coordinates is always positive

$$\left(\frac{\partial T}{\partial s}\right)_{x_i} > 0$$

and hence the directions of varying entropy and temperature on these curves are identical (Fig. 4.1).

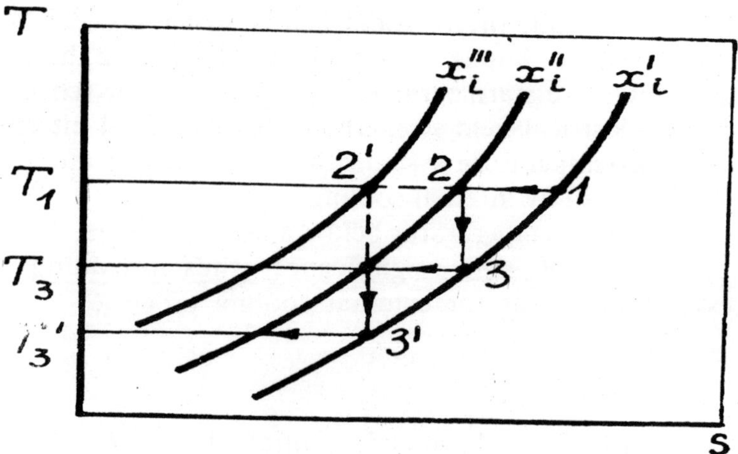

Fig. 4.1. General cooling principle on the T-s diagram.

In fact, decreasing the temperature reduces the thermal velocity of microparticles, which at x_i = **const** must have greater order in the system; as a consequence, the entropy, being a measure of the system disorder, will decrease. Thus, system temperature reduction may be achieved by decreasing entropy at an invariable parameter x_i.

The simplest method of decreasing the entropy at

x_i = **const**, thereby decreasing the temperature, consists in removing some amount of heat from a system (line 1-3 in Fig. 4.1 corresponds to this process). However, such a method represents external cooling and has no independent significance since its realization needs a lower-temperature system compared with the one considered.

The stated problem may be solved in another way if only the parameter x_i can easily be varied up and down. Transition of a system to state 3 with a temperature lower than that in initial state 1 is made in two stages.

In the first stage, under isothermal conditions, the parameter x_i is varied in the direction corresponding to a system entropy decrease (process 1-2 in Fig. 4.1). In this case, it is necessary to have a heat receiver capable of absorbing, heat removed from this system during this process, at a temperature T_1. Such a heat receiver may be represented either by an ambient medium (in this case, the initial system state temperature T_1 is equal to the ambient value T_0) or by any external system, with its temperature originally decreased to $T_1 < T_0$ by the described method. The entropy decrease during process 1-2 occurs because of reduction of that part of the entropy s_x which is determined by the chosen parameter while another of its parts s_T determined by the temperature remains invariable. At this stage, work is done with the system, and heat is removed: for a reversible process the amount

$$Q = T_1 (s_1 - s_2) .$$

In the second stage, the parameter x_i changes in the opposite direction but already unter the isentropic conditions (s = **const**, process 2-3 in Fig. 4.1). At the same time, the components of the entropies s_x and s_T change. The first increases, which at s = **const** is possible only while decreasing the second component, or the quantity s_T. As a result, the system temperature decreases up to T_3.

Thus, for internal cooling to occur, the entropy must depend, at least, on one further parameter in addition to temperature. Isothermally varying this parameter in the direction of the cause of the entropy decrease, followed by a change in the opposite direction under isentropic conditions reduces the system temperature. It should be emphasized that the first stage is preparatory: it is necessary only that the quantity x_i be isentropically changed at the second stage in the required direction, or in the direction yielding a temperature drop. If such a possibility exists in the initial state, then naturally there is no need in the first stage.

Obviously, for a lower finite temperature T_3' to be achieved, the parameter x_i must be varied over a wider range (a "step" in the **T-s** diagram 1-2'-3' in Fig. 4.1 formed by the processes **T = const** and **s = const** becomes broader). However, such a method always has li-mits since the entropy depends on the parameter x_i alone, and extending the x_i range may have technical limits. These difficulties may be avoided if use is made of one or several additional heat receivers with successively falling temperatures $T_{m1} > T_{m2} > ... > T_{mn}$; in this case, the temperature of the first T_{m1} is below the initial system state temperature T_1 while the temperature of the last one T_{mn} is higher than the temperature T_3' which must be attained (design of such heat receivers is not difficult).

With isothermal and isentropic processes proceeding over a relatively narrow range of x_i, a system may be cooled to a temperature level of the first of heat receivers T_{m1}, and then again by repeating these processes, the system temperature is decreased to a level of the next heat receiver, and so on.

The final process **s = const** which is carried out on the isothermal change of x_i at a temperature of the last heat receiver T_{mn} yields T_3' . On the **T-s** diagram (Fig. 4.1) such cascade cooling is shown by the broken line consisting of alternating horizontal **(T = const)** and vertical **(s = const)** sections.

Reduction of temperature by this will be observed unless the interaction energy **W** between individual system particles remains much lower than the thermal one **kT** (**k** = $1.38 \cdot 10^{-23}$ J/K is Boltzmann's constant). At temperatures $T \approx W/k$, this is violated and the system order starts failing spontaneously, when affected by the particle force interaction without interference (varying the parameter x_i under isothermal conditions is interference of this type). Although is very small, the quantity is finite, and so for any system and any parameter x_i it is always possible to indicate the temperature level below which the entropy stops depending on x_i and further temperature decrease becomes impossible.

This conclusion once more supports the validity of Nernst's law, implying that an absolute temperature zero cannot be attained.

The general cooling principle stated is, to a known extent, some-what of an idealization. As well as different "non-ideal facts" due to irreversibility losses, under real conditions it appears to be not always expedient to arrànge pure isothermal and pure isentropic processes, a result of the necessity to provide high performance characteristics of the equipment used, e.g. reliable operation, etc. As an example of this type of conscious deviation from an ideal process, in a number of cases isentropic expansion of a vapor-gas mixture in an expander is replaced by isoenthalpic expansion in a throttle.

4.2. Classification of cooling methods

Depending on the parameter to be varied, cooling methods may be subdivided into several groups. If we confine ourselves to considering only those methods which prove to be effective for obtaining cryogenic temperatures, then the latter may be classified as follows (the parameter to be varied is given in brackets):

1) thermomechanic methods of internal cooling (a parameter to be varied is pressure, $x_i = p$)

- isentropic expansion,
- adiabatic expansion,
- evaporation of acondensed phase with removal of the forming vapor,
- adiabatic throttling,
- desorption cooling;

2) magnetocaloric methods of internal cooling (a parameter to be varied is magnetic field intensity, $x_i = H$)

- Ettingshausen effect,
- adiabatic demagnetization,
- nuclear demagnetization,
- magnetization of superconductors;

3) electrocaloric methods of internal cooling (a parameter to be varied is electric field strength, $x_i = E$)

- Peltier effect,
- depolarization of dielectrics.

4) Methods that use the quantum properties of helium isotopes He^3 and He^4, as well as their solutions constitute a special group. This group will by convention be called thermoquantum methods of internal cooling. The methods belonging to this group, as well as to the first, are connected with varying the pressure ($x_i = p$):

- He^3 dilution in He^4,
- Pomeranchyuk effect,
- Kapitza's mechanocaloric effect.

4.3. Classification of temperature ranges

It is possible to distinguish three ranges of low temperatures, whose boundaries are determined by some convention:

1) moderately low temperature (300 K to 120 K) range covering air conditioning, food storage, chemical reactor cooling, etc.;

2) deep cold region, termed cryogenic, over a temperature range 120 K to 0.3 K. It covers separation of air and industrial gases, liquefaction of methane, oxygen, nitrogen, hydrogen, neon, helium and cooling of different objects by these fluids;

3) range of ultralow temperatures (below 0.3 K) used in physical experiments.

Temperatures in the first and second ranges are mainly obtained by thermomechanical processes. In addition, electro- and magnetocaloric processes are used.

Obtaining ultralow temperatures is connected with employing the quantum effects in helium isotopes, as well as the electro- and magnetocaloric effects.

4.4. Classification of systems by purpose

Cryogenic systems are subdivided by purpose into the following three groups.

1) Refrigerators are intended for cooling and thermostating different objects. As a rule, these problems are being solved simultaneously: an object is cooled to an assigned temperature level and is kept in this state using some cooling process.

As examples of refrigeration units we may cite cryogenic refrigerators for cooling and keeping a low temperature in research cryostats, infrared and electronic systems, deep vacuum chambers, superconducting electrical plants (generators, engines, cables, magnets, etc.), in biology and surgery, as well as in other fields of science and technology.

2) Liquefying plants are intended for gas liquefaction. Cryogenic liquefying plants are widely used to produce liquid methane, oxygen, nitrogen, hydrogen, helium.

3) Gas-distribution plants are intended for gas mixture separation into respective components. Separation requires gas mixture cooling to a phase transition temperature. Moreover, often a separation product is produced in liquid form. Usually, atmospheric air serves as the initial gas mixture; similar processes are also used for separating natural and by-product gases, for extracting rare isotops (deuterium from hydrogen, $\mathbf{He^3}$ from $\mathbf{He^4}$).

Questions

1. What is the difference between external and internal cooling?
2. What is the general principle of internal cooling of an arbitrary system?
3. What are the advantages of the cascade cooling process?
4. What force parameters can be varied to obtain cryo-temperatures?
5. Describe the main specific features of thermo-mechanical, magnetocaloric, electrocaloric, and thermoquantum cooling methods.
6. What are the specific features of each of the three low temperature ranges?
7. What are the specific features of refrigerating, liquefying, and gas-distribution systems?

5

THERMOMECHANICAL PROCESSES OF INTERNAL COOLING: THROTTLING AND EXPANDING

Thermomechanical processes of internal cooling include:

- throttling, i.e. pressure decrease of gas or fluid flow via a local hydraulic resistor (plug, valve, etc.) without heat exchange with the surroundings and without any external work;
- expanding, i.e. gas expansion accompanied by external work;
- exhaust, i.e. free gas discharge from a vessel;
- vapor pump-out, i.e. a fluid temperature decrease due to reduced vapor pressure above the fluid surface;
- desorption, i.e. a solid adsorbent temperature decrease due to vapor pump-out of adsorbent gas ;
- gas expansion in the vortex tube (Ranque-Hilsch tube), i.e. gas flow separation into cold (axial) and heat (peripheral) flows.

However, the vortex tube as a cold generator has low efficiency and will not be considered in detail.

Specific thermomechanical processes of cooling in helium isotopes associated with producting ultralow temperatures will be analyzed in a separate lecture.

Consider in more detail thermomechanical processes of cooling to be used for producing cryogenic temperatures.

5.1. Throttling

During throttling adiabatic gas expansion occurs under steady flow conditions, without external work and without changing velocity on the controlling surface. It should be emphasized that flow steadiness, first of all, assumes pressures to be constant before and after throttling, and the absence of flow velocity increase on the controlling surface does not exclude a local increase (or decrease) inside a system, for example in the throttle device itself.

To realize this process in practice some type of hydraulic resistor - throttle valve, demper, calibrated slit, etc. - is set along the gas path.

5.1.1. Thermodynamics of throttling

Joule and Thomson studied this process in the following manner (Fig. 5.1). The steady gas flow (initial temperature T_1) slowly passed through copper pipe (3) and then plug (2) protected by screens (1). In experiments, a temperature T_2 change was fixed at a varying pressure drop $\Delta p = p_1 - p_2$.

Consider an elementary volume of the slow gas flow in two sections located on both sides of the plug and at a sufficient distance from it, where the motion may be assumed steady. The energy conservation law for an open system with one transit steady gas flow is

$$q = l + h_2 - h_1 + \frac{w_2^2 - w_1^2}{2} + g\,(z_2 - z_1)\,, \qquad (5.1)$$

where q is heat per unit mass ($J \cdot kg^{-1}$); l is work; h is enthalpy; w is velocity; g is gravity acceleration; z is vertical coordinate.

Assuming that gas velocities in the considered sections are the same and the coordinate difference $(z_2 - z_1)$ is small, for the throttling process at $l = 0$ and $q = 0$ we have

$$h_2 = h_1 = \text{const} \qquad (5.2)$$

or

$$u_1 + p_1 v_1 = u_2 + p_2 v_2 = \text{const}\,,$$

where u is internal energy per unit mass; v is specific volume.

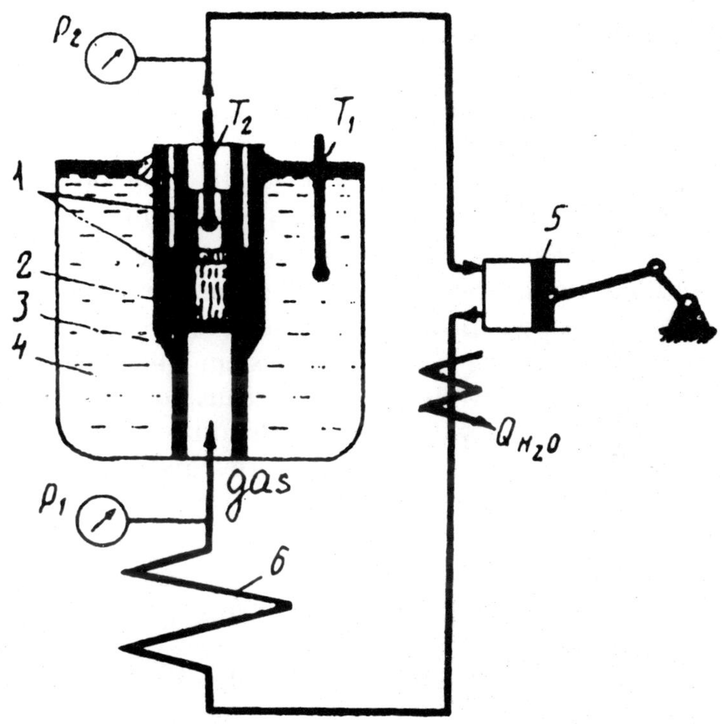

Fig. 5.1.Schematic of Joule and Thomson's device for studying gas thotting: 1, rubber ring screens; 2, cotton plug; 3, cooper pipe; 4, calorimeter; 5, compressor; 6, coil heat exchanger.

No assumptions of gas properties were made; therefore, the obtained result is valid for both ideal and real gases: in throttling of gas the enthalpy does not change.

For an ideal gas, the change in internal energy and enthalpy in any process is determined by:

$$\mathbf{du = c_v\ dT\ ; \qquad dh = c_p\ dT\ .} \tag{5.3}$$

Hence it follows that if **dh = 0**, then **dT** and **du** are equal to zero as well.

Thus, during throttling of the ideal gas

$$\begin{aligned} &\mathbf{dh = 0}\,; \qquad &&\mathbf{h = const}\,; \\ &\mathbf{dT = 0}\,; \qquad &&\mathbf{T = const}\,; \\ &\mathbf{du = 0}\,; \qquad &&\mathbf{u = const}\,. \end{aligned} \tag{5.4}$$

For a real gas, eqn. (5.2) does not yield the temperature and internal energy to be constant on throttling.

When a real gas expands, distances between molecules grow and increase is done against molecular attraction forces. Moreover, in real gas flow, the hydrodynamic force work of each unit mass at the controlling system inlet and outlet equal to a product **pv** is different because of the different compressibility. Values of these individual works predetermine a change in the internal energy and temperature; therefore, in the general case, for a real gas under throttling we obtain

$$\begin{aligned} &\mathbf{dh = 0}\,; \qquad &&\mathbf{h = const}\,; \\ &\mathbf{dT \geq 0} \quad \text{or} \qquad &&\mathbf{dT < 0}\,; \\ &\mathbf{du \geq 0} \quad \text{or} \qquad &&\mathbf{du < 0}\,. \end{aligned} \tag{5.5}$$

Thus, the real gas temperature on throttling may be decreased and increased.

A throttling process is irreversible, and pressurizing again is impossible without doing work.

Ideal gas throttling is completely irreversible since it is accompanied by no effects which could bring about a return to the initial state. Entropy increase is maximum and equals an entropy decrease under isothermal compression of an ideal gas.

The real gas throttling process is partially reversible as it is accompanied by changing temperature; in this case, a thermal reservoir is created with a higher or lower temperature, and, as a result, there is the possibility of using a temperature drop for doing work (this work may be used to return a gas to its initial state). The entropy increase during real gas throttling is not equal to the entropy change at isothermal gas compression.

Determine a real gas temperature change during throttling.

The determination of the enthalpy **h = u + pv** yields (with regard to the first thermodynamics law) an expression for a complete enthalpy differential in **s-p** variables:

$$dh = du + p\,dv + v\,dp = T\,ds + v\,dp. \qquad (5.6)$$

Hence,

$$\left(\frac{\partial h}{\partial p}\right)_T = T\left(\frac{\partial s}{\partial p}\right)_T + v \qquad (5.7)$$

or, with regard to the differential Maxwell eqn. (3.15), we have

$$\left(\frac{\partial h}{\partial p}\right)_T = -T\left(\frac{\partial v}{\partial T}\right)_p + v\,. \qquad (5.8)$$

In **T-p** coordinates, the complete enthalpy differential is of the form:

$$dh = \left(\frac{\partial h}{\partial T}\right)_p dT + \left(\frac{\partial h}{\partial p}\right)_T dp \qquad (5.9)$$

or, with regard to (5.8), we have

$$dh = c_p\,dT + \left[-T\left(\frac{\partial v}{\partial T}\right)_p + v\right] dp\,. \qquad (5.10)$$

Hence, for throttling **(dh = 0)** we obtain:

$$\alpha_h = \left(\frac{\partial T}{\partial p}\right)_h = \frac{1}{c_p}\left[T\left(\frac{\partial v}{\partial T}\right)_p - v\right] = \frac{vT}{c_p}\left(\beta - \frac{1}{T}\right), \qquad (5.11)$$

where

$$\beta = \frac{1}{v}\left(\frac{\partial v}{\partial T}\right)_p$$

is the temperature coefficient of volume expansion (isobaric compressibility).

The quantity

$$\alpha_h = \left(\frac{\partial T}{\partial p}\right)_h,$$

called the differential Joule-Thomson effect determines the temperature change for infinitesimal pressure drop during throttling. Eqn. (5.11) is valid for both gases and fluids. For an ideal gas, $\alpha_h = 0$ since in this case

$$\left(\frac{\partial v}{\partial T}\right)_p = \frac{v}{T}.$$

In practice, the throttling process is always characterized by a finite pressure difference; therefore, for such a process, integrating eqn. (5.11) yields:

$$T_2 - T_1 = \int_{p_1}^{p_2}\left(\frac{\partial T}{\partial p}\right)_h dp = \int_{p_1}^{p_2}\frac{1}{c_p}\left[T\left(\frac{\partial v}{\partial T}\right)_p - v\right]dp. \qquad (5.12)$$

This expression is responsible for the so-called integral Joule Thomson effect (temperature change for finite pressure difference).

In designing low-temperature plante, the integral throttling effect is usually determined by thermodynamic diagrams (Fig. 5.2) or special plots (Fig. 5.3) are constructed.

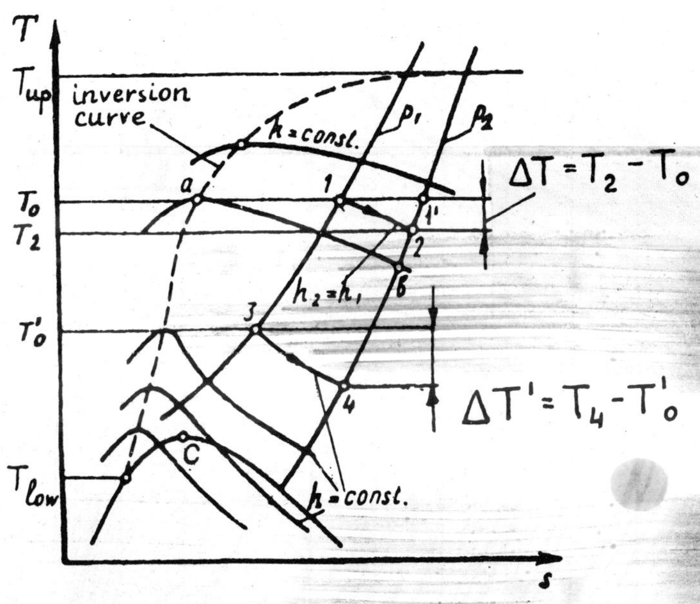

Fig. 5.2. Throtting process on the T-s diagram.

5.1.2. Inversion. Positive and negative throttle-effect

From eqns. (5.11) and (5.12) it follows that the sign of the throttle-effect may vary. If

$$\frac{\Delta T}{\Delta p} = \frac{T_2 - T_1}{p_2 - p_1} > 0 ,$$

then $T_2 < T_1$ (cooling takes place) since p_2 is always smaller than p_1 ; if $\Delta T / \Delta p < 0$, then $T_2 > T_1$, which corresponds

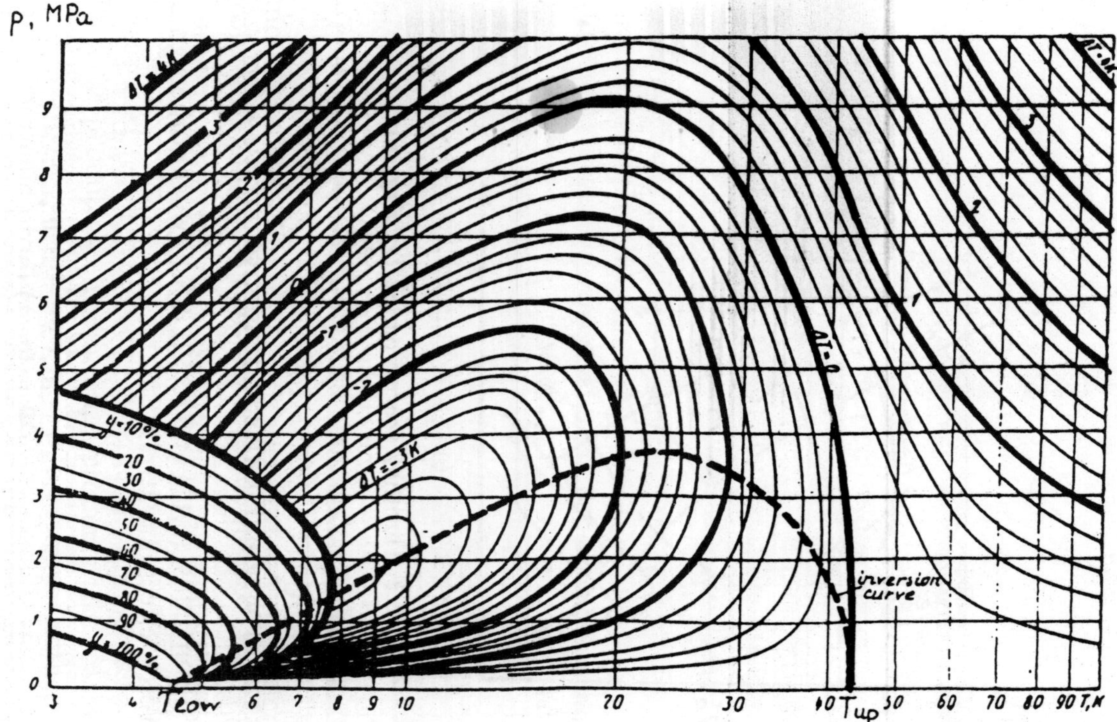

Fig. 5.3. Helium temperature change diagram on throttling up to a pressure of 0.1 MPa : p and T, pressure and temperature before throttling; y, amount of fluid after throttling.

to heating. A sign change in the throttle-effect is called inversion. At an inversion point,

$$\left(\frac{\partial T}{\partial p}\right)_h = 0.$$

States in which $(\partial T/\partial p)_h = 0$ are determined by the inversion curve eqn.

$$T\left(\frac{\partial v}{\partial T}\right)_p - v = 0 \tag{5.13}$$

or

$$\left(\frac{\partial v}{\partial T}\right)_p = \frac{v}{T}. \tag{5.14}$$

The inversion curve separates the regions of positive throttle-effect (cooling) and negative throttle-effect (heating). Fig. 5.4 shows plots of inversion curves for a number of gases. The region under the inversion curve corresponds to a positive throttle-effect. The eqn. for the inversion curve in relative (reduced) gas values obeying the Van der Waals eqn. is of the form

$$\pi = 24\sqrt{3\tau} - 12\tau - 27, \tag{5.15}$$

where $\pi = p / p_{cr}$; $\tau = T / T_{cr}$; p_{cr} and T_{cr} are parameters of critical point.

In accordance with Fig. 5.2 and Fig. 5.4, for each substance there exists a maximum inversion temperature T_{up} above which at any pressure the throttle-effect is negative. This temperature is called the upper inversion temperature. Also, there exists the so-called lower inversion temperature T_{low} in the fluid region. The corresponding point lies on the boundary curve (Fig. 5.2 and Fig. 5.4).

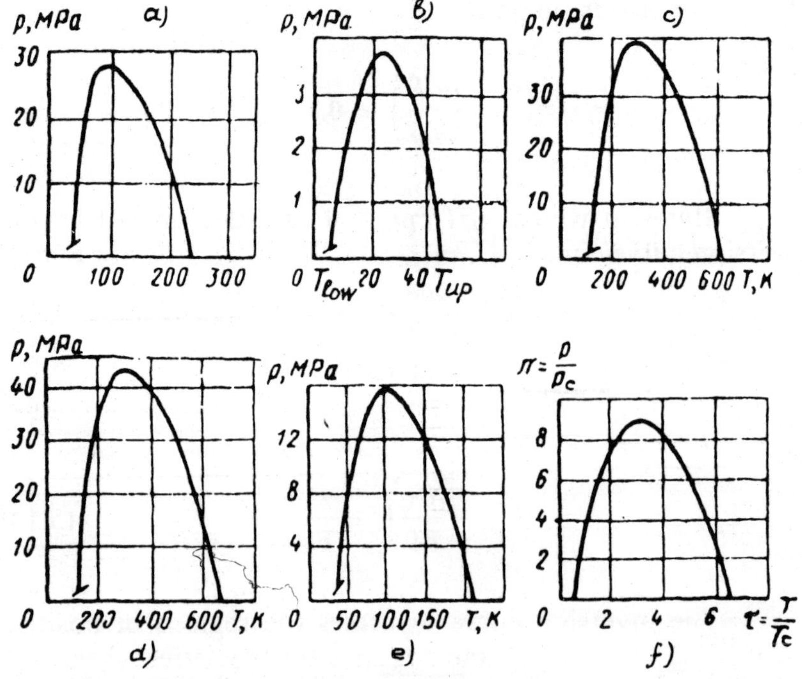

Fig. 5.4. Inversion curve for various gases: a, Ne; b, He; c, N_2; d, air; e, H_2; f, Van der Waals gas.

The inversion curve in **T-s** coordinates (Fig. 5.2, dashed line) passes through extremum values of isenthalpy (**h = const**) and asymptotically approaches T_{up}, where T_{up} is the inversion temperature at $p \to 0$. The inversion temperature varies writh the gas (Fig. 5.4). For some gases (air, nitrogen, oxygen) the upper inversion temperature is higher than the mean ambient temperature; for other gases (helium, neon, hydrogen) it is lower. As seen from Fig. 5.2, for a gas to be cooled under throttling, it is first necessary to decrease its temperature to a value lower than the upper inversion temperature T_{up}. If this condition is satisfied, then a question arises: up to what pressure p_1 should a gas be compressed to attain the maximum integral effect ΔT with know initial temperature T_0 (T_0 is the ambient temperature or preliminary cooling temperature). Differentiating eqn. (5.12) with respect to p and equating the obtained expression to zero, we have

$$\left(\frac{\partial T}{\partial p}\right)_h = 0. \tag{5.16}$$

This expression implies that an unknown point a in Fig. 5.2 lies on the inversion curve, and the value of the pressure of interest to us is equal to the inversion pressure at a given temperature. The above clearly follows from Fig. 5.2: on gas expansion from point "a" to point "b", the magnitude ΔT is maximum. However, in throttle cycles the pressure recommended in practice may be less than the inversion point pressures. For example, in the case of air compression (at $T = 300$ K, a pressure $p \approx 20$ MPa is used instead of 39 MPa). In helium and hydrogen throttle cycles, optimum compression pressures are close to inversion pressures.

For small pressures, the quantity α_h for a given gas in practice depends only on temperature. For instance, Joule and Thomson who carried ont experiments up to pressures of 0.6 MPa over a temperature range of 273÷373 K obtained the empirical relation

$$\alpha_h = a_0\left(\frac{273}{T}\right)^2, \tag{5.17}$$

where a_0 is a constant.

Thus, in the gaseous state region for a T decrease, the differential throttle-effect increases.

Increasing gas density, α_h starts to depend on pressure. Later experiments have revealed that α_h decreases with increasing pressure, and a plot of α_h versus p is close to linear:

$$\alpha_h = (a_0 - b_0 p)\left(\frac{273}{T}\right)^2, \qquad (5.18)$$

where b_0 is a constant.

In practice, the pressure influence on α_h is only at pressures of several megapascals.

The above implies that the initial temperature of the throttling process should be lowered to augment the integral effect. However, it should be clearly understood that in low-temperature throttle cycles, the refrigerating capacity of a theoretical cycle without cold loss will not depend on the initial throttling temperature.

Augmenting the integral throttle-effect by decreasing temperature opens up the possibility of increasing refrigerating capacity. In practice, this requires so-called preliminary cooling to be arranged.

The preliminary cooling role is reduced to creating a heat reservoir with a temperature T_0' lower than ambient temperature T_0. In the following, the compressed and non-compressed gas enthyalpy difference grows in the region of the positive throttle-effect at the same temperature, thereby elevating the refrigerating capacity.

This enthalpy difference, often adopted in engineering calculations, is denoted by Δh_T in the general case and is called the thermal throttle-effect (or sometime the isothermal throttle-effect).

The quantity Δh_T is found by thermal diagrams as the compressed and non-compressed gas enthalpy difference at a given temperature, e.g. at $T = T_0$ (Fig. 5.2):

$$(\Delta h)_{T0} = h_1' - h_1 = h_1' - h_2. \qquad (5.19)$$

Consider some physical reasons for temperature change and the existence of inversion under throttling. From eqns. (5.8) and (5.11) α_h is found:

$$\alpha_h = \left(\frac{\partial T}{\partial p}\right)_h = \frac{-1}{c_p}\left(\frac{\partial h}{\partial p}\right)_T \quad (5.20)$$

hence

$$\Delta T = T_2 - T_1 \approx \frac{h_1 - h_1'}{c_p}. \quad (5.21)$$

Here $(\partial h/\partial p)_T$ is the enthalpy change in the elementary process of isothermal compression.

Eqn. (5.21) yields a more exact result if c_p is determined as the mean value for pressure at the end of the expansion process over a temperature range $T_1 \div T_2$.

5.2. Expanding

Analysis of the expanding process will begin by considering isentropic gas expansion.

Gas expansion under adiabatic conditions, i.e. no external heat exchange may occur without changing the entropy in the absence of any internal friction processes. As a result, all possible energy of a compressed gas must be converted into external work in order for the condition $s = const$ to be satisfied. It is evident that in this case, the internal gas energy decrease is maximum (as against other expansion processes with the same initial parameters and degree of expansion); therefore, such a process is acompanied by the largest temperature decrease. It is essential that the work done by a gas in this process be completely transferred to a device iso-lated from the gas.

5.2.1. Thermodynamics of expanding

The magnitude of the work done by a gas in closed and open systems at $s = const$ is different:

$$l_{closed} = u_1 - u_{2s} = \int_{\circ} p\,dv\ ; \tag{5.22}$$

$$l_{open} = h_1 - h_{2s} = \int v\,dp\ . \tag{5.23}$$

The latter is valid for the same gas flow velocities at inlet and outlet. Since real processes of flow and gas expansion cannot occur without friction, under adiabatic conditions the process **s = const** cannot be realized in practice. This process is ideal. Nevertheless, its analysis is of substantial importance for comparing efficiencies of throttling and expansion processes.

To find temperature change during isoentropic expansion, use is made of an expression for the entropy differential in **T-p** variables

$$ds = \left(\frac{\partial s}{\partial T}\right)_p dT + \left(\frac{\partial s}{\partial p}\right)_T dp = \frac{c_p}{T} dT + \left(\frac{\partial s}{\partial p}\right)_T dp\ . \tag{5.24}$$

Hence, at **s = const (ds = 0)**, for both an open and a closed system a differential expanding effect

$$\alpha_s = \left(\frac{\partial T}{\partial p}\right)_s = \frac{-T}{c_p}\left(\frac{\partial s}{\partial p}\right)_T \tag{5.25}$$

or, with regard to the differential Maxwell eqn. (3.15), we have:

$$\alpha_s = \frac{T}{c_p}\left(\frac{\partial v}{\partial T}\right)_p = \frac{T}{c_p}\beta v\ . \tag{5.26}$$

5.2.2. Analyse of differential expanding effect formula

Comparing eqns. (5.11) and (5.26) yields:

$$\alpha_s = \alpha_h + \frac{v}{c_p} \tag{5.27}$$

and

$$\frac{\alpha_h}{\alpha_s} = 1 - \frac{1}{\beta T} \tag{5.28}$$

On the basis of the relations obtained, the following may be concluded.

1. The quantity α_s is positive in practice in any region of working body states allowing expansion on physical grounds (exceptions are not considered here).

2. With increasing temperature, α_s grows, thereby increasing expansion work.

3. With increasing pressure, i.e. decreasing specific volumes with increasing working body density, α_s decreases. Thus, during $s = const$ expansion, the quantity α_s is variable.

4. In the vicinity of critical state and in the region of boiling fluid states $(c_p \to \infty)$, values of α_s and α_h are closest.

5. Relationships between α_s and α_h depend on the kind of gases and gaseous mixtures. For example, for methane in the region of about 293 K and about 6 MPa, the ratio $\alpha_h / \alpha_s = 0.5215$; for air in the region of the same temperatures and pressures $\alpha_h / \alpha_s = 0.1835$. The larger the ratio α_h / α_s, the less profitable in the general case is the use of expanders.

For an ideal gas $\alpha_h = 0$ and $\alpha_s = v/c_p$. In this case, isoentropic expansion is described by Poisson's eqn.

$$pv^{\gamma} = const \tag{5.29}$$

(γ is the adiabat index), and the relationship between a temperature and pressure is of the from:

$$\frac{T_1}{T_2} = \left(\frac{p_1}{p_2}\right)^{1-1/\gamma} . \tag{5.30}$$

and a temperature decrease is determined from the formula:

$$-\Delta T = T_1 - T_2 = T_1\left[1 - \left(\frac{p_2}{p_1}\right)^{1-1/\gamma}\right] . \tag{5.31}$$

where subscripts 1 and 2 denote parameters at the start and end of a process, respectively.

The quantity α_s for a real gas may be larger or smaller than for an ideal gas, depending on the sign of α_h . However, for the majority of real gases used in cryogenic engineering (except hydrogen), temperature change in isoentropic processes may be calculated through formula (5.31) with an accuracy sufficient for practical calculations. It is essential that eqn. (5.31) is valid for these real gases over a wide range of states, including those near the boundary curve. Hydrogen is an exception.

For hellium, neon, nitrogen, air the probable error in determining temperatures using formula (5.31) does not exceed 3%, but for hydrogen at a constant value of $\gamma = 1.407$ this may reach 20 ÷ 30% (at temperatures below 130 K). On physical grounds, this specific feature of hydrogen is attributed to the fact that over a temperature range 80 ÷ 250 K, the heat capacity and Poisson's index γ depend strongly on temperature. Recall that for hydrogen the characteristic temperature T_{rot} (degeneration temperature of a rotational molecule contribution to heat capacity) is high, about 80 K; therefore, at temperatures close to 80 K and lower, the heat capacity c_v for hydrogen becomes almost equal for c_v of a monoatomic gas.

To avoid great errors in calculating a hydrogen temperature change through eqn. (5.31), values of Poisson's index must be chosen as mean ones for $s = const.$ In this case, the required accuracy is attained.

In practice, gas expansion processes in doing external work are being perfomed in different expansion machines known as expanders. In expanders, the compressed gas energy is converted into work, and the process to some extent approaches isoentropic.

5.3. Comparison of throttling and expanding

It is of interest to compare the temperature effects to be attained during isenthalpy and isentropy expansion. From relation (5.27), it follows that

$$\left(\frac{\partial T}{\partial p}\right)_s - \left(\frac{\partial T}{\partial p}\right)_h = \frac{v}{c_p}. \tag{5.32}$$

As $v > 0$ and $c_p > 0$, we have $(\partial T/\partial p)_s > (\partial T/\partial p)_h$. So, as regards the thermodynamic approach, isentropic expansion is always preferted for decreasing a temperature over adiabatic throttling.

Their difference increases with rise in specific volume v, i.e. with elevating temperature or reducing pressure However, in the region near the critical point where the hea capacity c_p rises sharply, throttling becomes almost a effective a process in the sense of cooling as isentropic expan sion.

Moreover, the temperature effects of both means hav in general roved to be, the same for pure substances in a two phase region when $c_p \to \infty$ (Fig. 5.5a). Even in this case, no wever, the enthalpy of a working body after isentropic expansion will be smaller than after throttling, Therefore, although the obtained temperatures are also the same, the amount of "cold" to be produced for **s = const** is greater. For mixtures, the relation $(\partial T/\partial p)_s > (\partial T/\partial p)_h$ is also valid in a two-phase region (Fig. 5.5b).

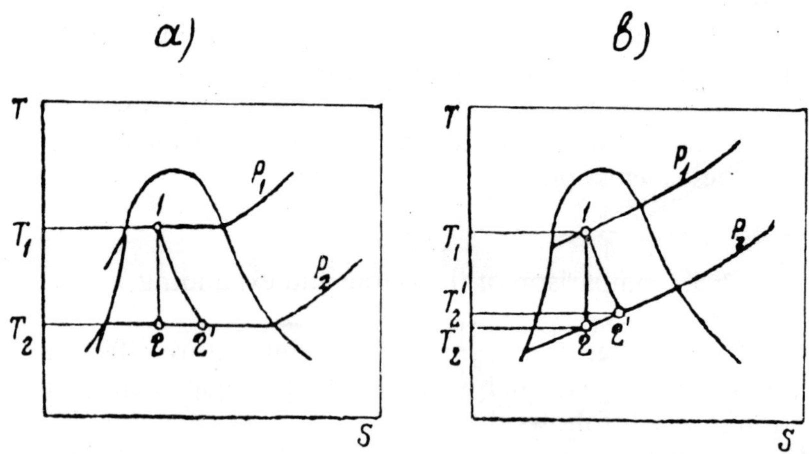

Fig. 5.5. Isentropic (1-2) and isenthalpic (1-2') expansion in in a two-phase region: a, pure substances; b, mixtures.

One more clear advantage of isentropic expansion is hat in this process, temperature is always decreased, and loes not depend on inital parameters of a system. At the ame time, use of throttling as a cooling means is restricted by the inversion curve.

Still, despite the disadvantages mentioned, enthalpy expansion is completed successfully by isentropic expansion n cryogenic engineering. This is attributed to a simple design, and, the consequent highly reliable operation of valves as throttle devices.

Questions

1. What is meant by the throttling process?

2. What is the differential Joule-Thomson effect α_h called?

3. Derive the formula for calculation of α_s.

4. What is the difference between the positive end negative throttle-effect?

5. Derive the inversion equation.

6. Compare the differential effects of throttling and expanding.

7. How does α_s depend on gas temperature and pressure?

8. How is the process of isoentropic expansion for an ideal gas calcucated?

9. What are the advantages and disadvantages of the throttling and expanding processes?

6

THERMOMECHANICAL PROCESSES OF INTERNAL COOLING: EXHAUST, VAPOR PUMP-OUT AND DESORPTION

In the last chapter we considered the production of "cold" means of throttling and expanding. Now we continue our study of the thermomechanical processes of cooling to be used for production low temperatures.

Consider following thermomechanical processes:

- exhaust;
- vapor pump-out;
- desorption.

6.1. Exhaust

Exhaust is free discharge of a gas from a vessel. Consider adiabatic expansion of a gas to be discharged from some volume: cylinder, sphere, etc. This process is one of the most widely used. It is very often used in low-temperature gas machines. The scheme of an exhaust process is very simple. A compressed-gas cylinder has an escape valve; with it open the gas quickly discharges from the cylinder and moves to the pipeline (Fig. 6.1).

Exhaust process is unsteady and non-equilibrium adiabatic expansion of a gas whilst doing external work. Heat exchange between the gas and walls is excluded according to the condition, and non-equilibrium is caused by the fact that, on varying the volume, the gas pressure acting on the controlling surface of a system is not balanced by counterpressure forces. To analyze this process it is convenient to use the scheme in Fig. 6.1. Initial cylinder gas parameters are $\mathbf{T}_i$ and $\mathbf{p}_i$. The cylinder is sealed with a valve-gate. After the gate is freed it starts moving with no

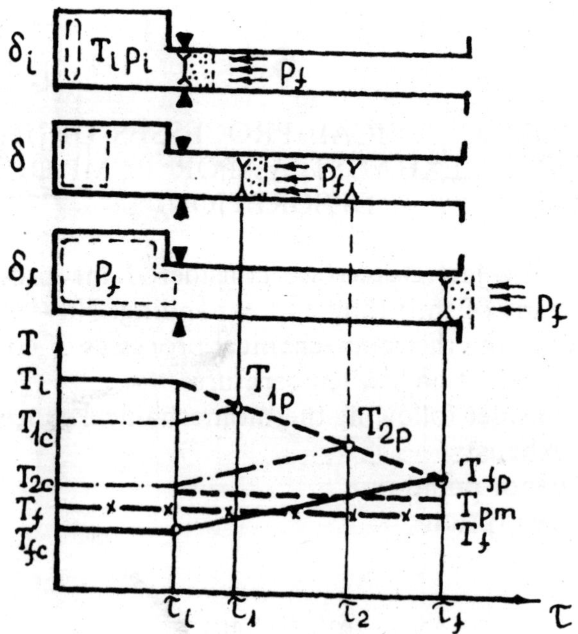

Fig.6.1. Scheme for realizing an exhaust process; conventional diagrams of instantaneous and equilibrium temperature distributions.

friction in an exhaust pipe.

The gas exerts on the gate a pressure which gradually drops. On other hand, a constant force of a counterpressure acts upon the gas because the gas flows into a region of constant pressure p_f. When the cylinder pressure attains a value p_f, the gate will stop moving. This will occur at a time τ_f. At some time after ceasing the process, an equilibrium temperature T_f is set in the system. T_f may be calculated as follows. For this non-equilibrium adiabatic process (per 1 kg of gas),

$$u_f - u_i \approx -p_f (v_f - v_i) . \qquad (6.1)$$

If the heat capacity c_v and compressibility coefficient $Z = pv / (R T)$ are considered constant, then we have

$$u_i - u_f = c_v(T_i - T_f) = p_f R Z\left(\frac{T_f}{p_f} - \frac{T_i}{p_i}\right). \qquad (6.2)$$

Recalling that at Z = **const**, $R\,Z\,/\,c_v = \gamma - 1$, from eqn. (6.2) we obtain

$$\frac{T_i}{T_f} = \frac{\gamma}{1 + (\gamma - 1)p_f / p_i} \qquad (6.3)$$

and

$$T_i - T_f = T_i\frac{\gamma - 1}{\gamma}\left(1 - \frac{p_f}{p_i}\right). \qquad (6.4)$$

A gas temperature decrease for the same T_i, p_i, and p_f in an non-equilibrium adiabatic process is smaller than for the isoentropic case.

It should be borne in mind that at $\Delta p = (p_i - p_f) \to dp$, quasi-equilibrium conditions are obeyed. In this case, the differential effect of exhaust and that of isoentropic processes are almost the same:

$$\alpha_{exp} = \frac{T_i}{p_i}\frac{\gamma - 1}{\gamma} = (\alpha_s)_i \qquad (6.5)$$

Thus, the mathematical form of the eqn. for the exhaust process is α_{exh} = **const**. Taking into account the relationship between α_{exh} and α_s, it may be assumed that the accuracy in calculating real gas (except hydrogen) temperatures during exhaust by eqns. (6.3) and (6.4) obtained for constant Z and γ is quite satisfactory. When calculating temperatures in the equilibrium adiabatic process for hydrogen, use should be made of mean values of Poisson's index.

Based on eqn. (6.5), it is easy to plot the lines

$$\alpha_{exh} = (\alpha_s)_i = const$$

for the exhaust process in T-s coordinates for given p_i and T_i (Fig. 6.2). From this figure it is seen particularly that in using the exhaust process as a cold production process, high degrees of expansion in one step are irrational. The smaller the expansion degree, the higher the exhaust process efficiency.

The gas enthalpy change during exhaust may be found from eqn. (6.2), considering that $h = u + p\,v$:

$$h_i - h_f = ZRT_i\left(1 - \frac{p_f}{p_i}\right). \qquad (6.6)$$

Expression (6.6) determines the refrigerating capacity of the exhaust process.

If the exhaust process is considered in time, then it is essential to allow for a "temperature scatter" or a temperature gradient onset in the gas flow discharging from the cylinder

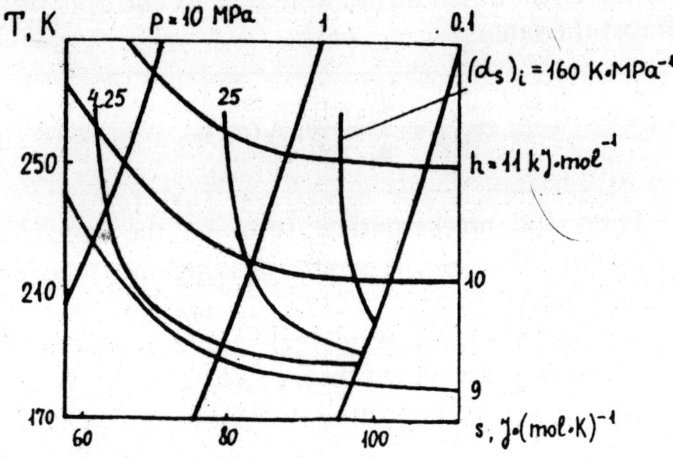

Fig. 6.2. Lines $(\alpha_s)_i$ = const for the exhaust process for the T-s diagram for air.

(Fig. 6.1). During the whole process at time instants τ_1, τ_2, τ_f, exhaust pipe temperatures T_1, T_{2p}, T_{fp} are higher than cylinder gas values T_{1c}, T_{2c}, T_{fc}. A cylinder gas temperature T_{fc} at the end of the exhaust process may be found, to a first approximation, by eqn. (5.30), as for the equilibrium adiabatic process. This is correct because on the controlling surface (contour δ in Fig. 6.1) of a subsystem containing a gas mass left in the cylinder at the end of the exhaust, the quasi-equilibrium conditions are considered to be satisfied. The mean temperature of gas to be discharged to the end of the exhaust T_{pm} may be simply calculated from the condition implying that after complete mixing of all gas portions left and discharges from the cylinder, their temperature must be equal to T_f. Thus, under the assumptions made we have

$$T_{pm} = \frac{T_i}{\gamma} \frac{p_i - p_f}{p_i - p_f (p_i / p_f)^{1-1/\gamma}} . \qquad (6.7)$$

L. Calete was the first who, in 1877, utilized the exhaust process to liquefy oxygen and other gases. In subsequent years it was used by E. Olshevsky and S. Vroblevsky to produce liquefied gases. In 1932, F. Simon successfully used this process for helium liquefaction. In 1959, U. Gifford and Mc. Magon designed an original cryogenerator where the exhaust process was repeated cyclically. The free gas discharge process was also utilized in the so-called pulse-tube.

Finally, it should be mentioned that the exhaust process is one of the most important processes of piston expanders.

6.2. Vapor pump-out cooling

The problem of stabilizing a temperature of different objects in cryogenic engineering is often solved by cryofluids, or fluids whose normal boiling point (p = 760 mm Hg) is below 120 K. In the simplest case, an object to be cooled is submerged in a volume of such a fluid which is normally in the saturation state because of inevitable environmental heat fluxes. Despite its triviality, this cryostating method has a number of advantages:

- it gives good thermal contact between the fluid and the object to be submerged in it ;
- parasitic heat flux does not affect the fluid saturation temperature and causes only fluid evaporation;
- a fairly uniform volume temperature distribution is set which may always be improved by forced fluid mixing;
- high reliability is achieved in the presence of cryofluid stored, necessary for reseach;
- last, the advantage of most interest: the fluid temperature may be varied over sufficiently wide ranges.

For pure substances on the equilibrium fluid-vapor line, the temperature is uniquely determined by pressure (Fig. 6.3). Varying it may yield any temperature over a range from T_{tp} (triple point) to T_c (critical point).

As shown in Fig. 6.3 the triple point pressure of all cryofluids is below atmospheric. Therefore, to lower the temperature of a sealed-cylinder fluid, compared to the normal boiling point T_{nb}, its vapors must be evacuated by vacuum pump. Elevating vapor pressure may increase the temperature. It should be emphasized that for this, it is not essential to use a compressor. In a sealed closed cryostat, the vapor pressure will increase because of evaporating fluid which, in turn, is caused by whatever existing environmental heat fluxes.

The rate of increasing pressure and, hence, temperature may be increased by placing extra heat sources in the cylinder.

To a first approximation, the vapor pump-out process may be interpreted as isentropic expansion if environmental heat fluxes are neglected and the boundaries of the considered system are located so that the vapor to be pumped down would be included in it. It may be assumed, for

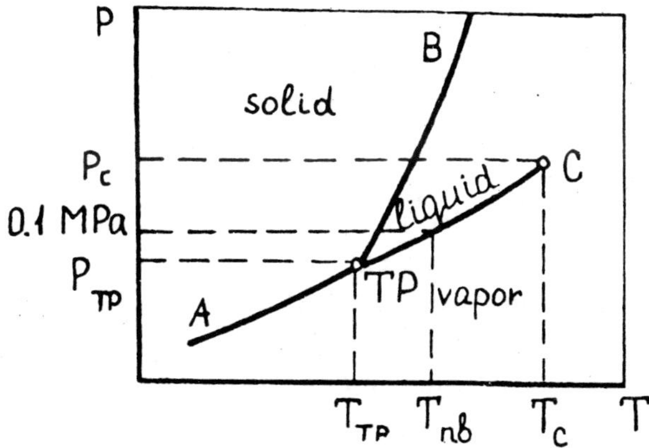

Fig. 6.3. Characteristic p-T diagram for substance: TP-C, boiling-condensation curve (saturation line); TP-B, melting-crystallization curve; TP-A, sublimation-desublimation curve.

example, that in the cylinder the fluid below the piston is in equilibrium with its vapors and the pressure above the piston is lowered by a vacuum pump.

As a whole obeying the general specific laws of the above process, isentropic expansion in a two-phase fluid-vapor system possesses some specific features that are to be analyzed. First, it is necessary to explain, on molecular level, the change in temperature of fluid and vapor to be at eqilibrium and, second, to evaluate this relationship quantitatively.

When equilibrium between the fluid and vapor is said to exist, this is understood to mean that the amount of molecules transiting from fluid to vapor per unit time exactly equal that to be returned to the fluid in this time, i.e. equilibrium should be considered dynamic (as the equality of zero-different rates of two opposite processes: evaporation and condensation). To leave the fluid, evaporating molecules must overcome a certain energy barrier that consists in work against the fluid molecule attractive forces, and in work against the external pressure of the already formed vapor. It is clear that all this work must be done only through the kinetic energy of the thermal molecule motion and only by those molecules that possess enough kinetic energy to do it. The rate of the reserve transition of molecules from vapor to fluid is proportional to vapor density.

It is now easy to imagine what will occur if in an equilibrium fluid-vapor system with initial pressure p_0 and temperature T_0, the pressure is adiabatically decreased by a quantity Δp (start of vapor pump-out). As the vapor pressure is decreased, the condensation rate also falls to a value determined by a new pressure value

$$p_1 = p_0 - \Delta p .$$

At the same time, because of decreasing of molecule work done to leave the fluid, molecules with smaller energy can enter the vapor, the evaporation rate will increase and, as a result, the equilibrium is upset.

However, a number of molecules capable of breaking away from the liquid decreases and the vaporization rate falls while the condensation rate is constant. It is clear that the loss of the fastest molecules results in decrease in fluid temperature. Finally, rates of both processes are equalized, and a new equilibrium state occurs, characterized by smaller values of pressure p_1 and temperature T_1. Similarly, it is possible to see why fluid temperature rises and how the equilibrium is brought about with increasing pressure.

Let us consider quantitative estimates. The relationship between temperature T and pressure p on the curve for any phase transition is established by the Clapeyron-Clausius equation. When applied to a fluid-vapor transition, this has the form

$$\left(\frac{dT}{dp}\right)_{sat} = \frac{T(v''-v')}{r} , \qquad (6.8)$$

where v'', v' are the specific volumes of vapor and fluid, respectively, and r is the specific vaporization heat.

It should be emphasized that unlike the isentropic and isenthalpy expansions considered earlier, the derivative dT / dp of interest to us is written, in this case, in complete differentials since a temperature on the saturation line is uniquely determined by a pressure. In addition, eqn. (6.8) supports a qualitative conclusion about the same direction in chànging: pressure and temperature on the saturation line. Indeed, the vapor density is always less than the fluid value and, therefore, $v'' > v'$. As a result, $(dT / dp)_{sat} > 0$, i.e. with decreasing pressure, the temperature is lowered as well, and vice versa.

To integrate eqn. (6.8), let us confine ourselves to pressures far from critical. In this region, $v'' >> v'$ and, moreover, the vapor may be, to a good approximation, interpreted as an ideal gas, which enables its specific volume v'' to be relate to temperature and pressure by the Clapeyron-Mendeleev eqn. $v'' = R T / p$, where R is specific gas constant, $J \cdot (kg \cdot K)^{-1}$.

Substituting this expression for the quantity v'' into eqn. (6.8) and neglecting in it the quantity v' yields

$$\left(\frac{dT}{dp}\right)_{sat} = \frac{RT^2}{r\,p}, \qquad (6.9)$$

At low pressures, the vaporization heat r depends slightly on temperature. So, for example, the quantity r for oxygen is lowered by 15% on decreasing temperature from normal boiling point to the triple point; such a change for nitrogen is 7%. Assuming, in relation (6.9), $r = const$ and integrating yields:

$$\ln p = \frac{-r}{R T_0} + C .$$

In order to determine the integration constant C, the

value of saturation pressure at some fixed temperature must be known. If $p = p_0$ when $T_s = T_{so}$, we have

$$\ln\frac{p}{p_0} = \frac{r}{R}\left(\frac{1}{T_{so}} - \frac{1}{T_s}\right). \qquad (6.10)$$

In principle, the choice of temperature T_{so} is quite arbitrary. Usually, the normal boiling point is used as this temperature.

It is quite straightforward to see that the relationship between $\ln(p/p_0)$ and $1/T_s$ appears to be linear (Fig. 6.4). This conclusion is consistent with the data obtained for different substances at pressures that are not high.

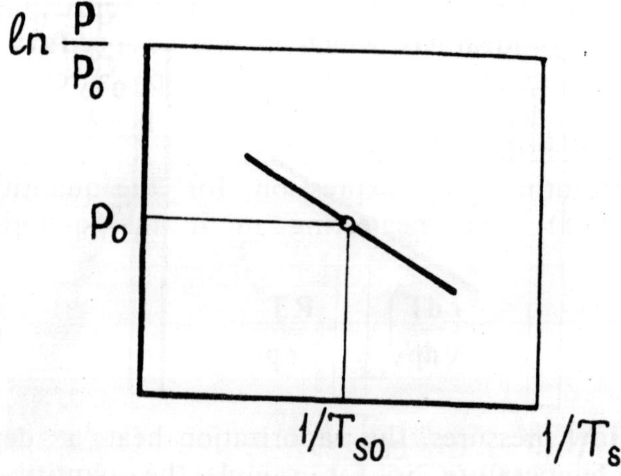

Fig. 6.4. Relationship between a temperature and pressure on the saturation line.

It is interesting that, as experimental findings illustrate, the relation predicted by eqn. (6.10) is also valid for many substances in the high pressure region where the quantity **r** depends strongly on temperature; the vapor certainly differs from an ideal gas and neglecting the quantity

v' as against v'' becomes inapplicable. Nevertheless, imposing these errors does not result in the relatively noticeable total effect, and the relation $\ln(p/p_0) \sim 1/T_s$ remains valid. Along with the dependence $T_s(p)$, when pumping out vapors it is advisable to know the fluid amount ΔG to be removed as vapor to lower fluid temperature by ΔT.

As already mentioned, under adiabatic conditions the work to change fluid into vapor with decreasing pressure results from the internal fluid energy. Considering a change from one state to another of an elementary substance portion dG accompanied by lowering fluid temperature by dT, we can write

$$r\, dG = c_s\, G\, dT\,, \tag{6.11}$$

where c_s is the specific fluid heat capacity on the saturation line.

Separating the variables and integrating eqn. (6.11) from some initial (1) to final state (2), we obtain

$$\ln\frac{G_2}{G_1} = \int_{T_{S1}}^{T_{S2}} \frac{c_s}{r}\, dT\,.$$

Here it is again necessary to refer to the experimental findings showing that for many cryofluids (nitrogen, oxyden, neon, para hydrogen) over a temperature range from normal boiling to triple point, the heat capacity c_s remains practically constant.

This temperature range conforms to the low pressure region where, as seen, the quantity r also remains at its constant value.

Thus, approximately it may be considered that

$$\ln\frac{G_2}{G_1} = \frac{c_s}{r}(T_{s2} - T_{s1})\,. \tag{6.12}$$

Hence, for the fluid flow-rate ΔG we have

$$\Delta G = G_1 - G_2 = G_1 \left\{ 1 - \exp\left[\frac{-c_s (T_{s1} - T_{s2})}{r} \right] \right\}. \qquad (6.13)$$

Comparison of relation (6.13) with experimental data on nitrogen and para hydrogen (Fig. 6.5) shows that it is quite applicable for practical estimates of flow-rate of fluid to be pumped out in order to cool. Calculations must use mean values of c_v and r over a temperature range $T_{tp} < T < T_{nb}$.

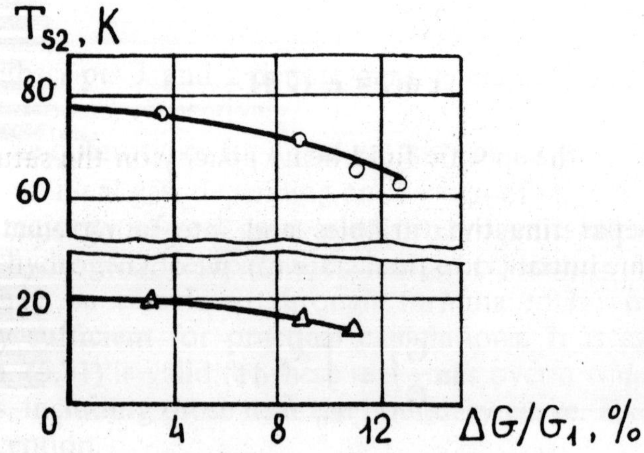

Fig. 6.5. Flow-rate of fluide to be cooled by vapor pump-out. Experimental data: o, nitrogen, T=77 K; Δ, hydrogen, T_{s1}=20 K; solid lines, calculation by eq. (6.13).

Note that in real systems when environmental heat fluxes inevitably exist the fluid flow-rate proves to be somewhat greater that follows from relation (6.13). Temperature range obtained using the cryofluids by the vapor pump-out method are limited by the temperature range of a liquid phase from triple to critical point.

Fig. 6.6 shows a plot of temperature regions where the commonest cryofluids exist. This figure illustrates the regions of the so-called helium, hydrogen, and nitrogen temperatures. It is easy to see that there are two temperature ranges: (1) from the triple oxyden point of 54.36 K to the critical neon point of 44.4 K; (2) from a the triple hydrogen point of 13.95 K to the critical helium-4 point of 5.2 K, over which no pure substance can exist in liquid state. This, of course, produces some difficulties of operation in these regions but they may be overcome completely (e.g. by using mixtures in a liquid state over the mentioned temperature ranges).

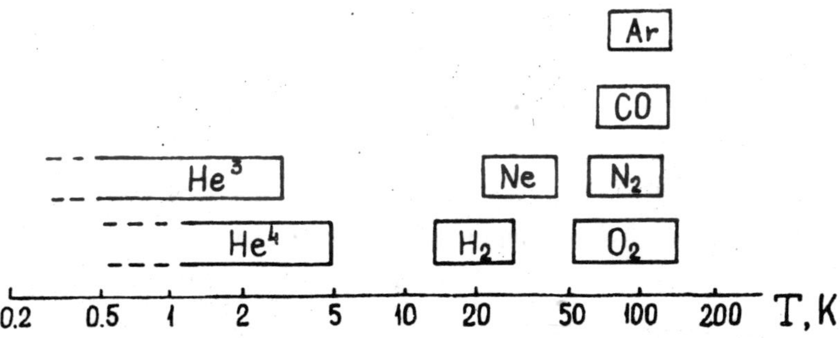

Fig. 6.6. Temperature regions of the existence of some cryofluids.

6.3. Desorption cooling

Desorption cooling is a relatively simple means of producing temperatures 5.2 to 13.95 K. Simon was the first to propose this method and put it into practice (Fig. 6.7).

Activated coal or some other adsorbent (silica gel, zeolite) in container 5 is cooled to possibly lower hydrogen temperatures.

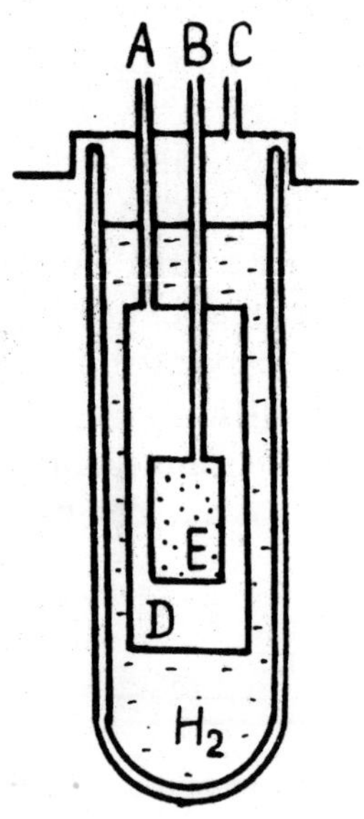

Fig. 6.7. Schematic of the device for desorption cooling by Simon's method. A,B, He pipes; C, H_2 pipe; D, casing; E, adsorbent.

To do this, hydrogen vapors are pumped out via pipe 3 from cryostat 6, and thermal contact between a hydrogen bath and adsorbent is provided by filling casing 4 with gaseous helium supplied via pipe 5. Then, helium is supplied via pipe 2 to container 5 where it is adsorbed by activated coal. Heat to be released under adsorption is removed to liquid hydrogen. Upon saturating the activated coal, the casing is pumped down to high vacuum, thereby destroying the thermal contact of container 5 with liquid hydrogen. If adsorbed helium is now pumped down, the coal temperature decreases. Changing the amount of desorbed helium may yield any temperature in the temperature range 5.2 ÷ 13.95 K and even penetrate into the helium temperature range (experiments have bean carried out where such a method was successful in lowering adsorbent temperature by 1.6 K).

Thus, the applicability of desorption cooling is confined to the hydrogen-helium range. This method may be used successfully at both lower and higher temperatures.

This method is, in the main, similar to isentropic expansion. Helium saturation of adsorbent is equivalent to isothermal gas compression (entropy of system decreasing), and adsorbed gas pump-out to expansion (provided that $\mathbf{s = const}$ the temperature decreases since an increase in that part of the entropy which is related to adsorption occurs).

It slould be emphasized that the temperature scale in Fig. 6.6 does not end with the value of 0.3 K by chance although, as is known, the helium triple point is absent and helium remains in liquid state up to absolute zero. In principle, vapor pump-out might lower the liquid helium temperature to a temperature as close as desired to absolute zero. However, in reality it is not simple to create the required rarefaction.

The degree of rarefaction that can be attained is mainly determined by two factors: perfection and capacity of the pumps, as well as the reliability of thermal insulation. The second factor is of special importance for operation with helium, since helium has a very small vaporization heat (at $\mathbf{T} = 4.2$ K, $\mathbf{r} \doteq 20$ kJ$\cdot$ kg^{-1}, 10 times less than that of nitrogen and 100 times less than that of water).

Even quite small heat fluxes result in the evaporation of large amounts of helium, the flow-rate of vapor to be pumped down increases, thereby causing deterioration of the vacuum for a given pump power.

One more property of helium proves to be extremely unfavourable: the very strong dependence of saturation temperature on pressure at $\mathbf{p} < 0.1$ MPa (Table 6.1).

For these reasons, liquid He^4 temperature may be lowered only to 1 K in real systems because of vapor pump-out. As regards record temperatures, in 1922 Kammerlingh-Onnes attained a temperature of 0.81 K when trying to produce the helium triple point. Ten years later, Keez managed to achieve a new record: 0.720 K when he used a set of powerful oil pumps.

Table 6.1. Helium isotope saturation temperature at various pressures.

T_s, K		2.0	1.0	0.8	0.5	0.3
p, mm Hg	He^4	2.767	0.120	11.45×10^{-2}	16.34×10^{-6}	3.88×10^{-10}
	He^3	150.55	8.564	2.744	0.1418	0.0015

After the Second World War, when as a result of the development of nuclear physics the refrigeration production of light He^3 isotope became accessible in sufficient amounts, the limit moved down several tenths of a degree to approx. 0.3 K. This value is the limit of the power of thermomechanical means. Very different means are needed to extend below this temperature.

Questions

1. How does the exhaust process occur?
2. Derive the equation for the exhaust process.
3. How is the refrigerating capacity if the exhaust process calculated?
4. What are the advantages of cooling using the gas pump-out method?
5. Explain, on the basis of molecular-kinetic theory, the process of reaching equilibrium writh pressure variation in a two-phase system.
6. How are fluid losses due to decreasing fluid temperature with gas pump-out calculated?
7. What cryotemperature ranges cannot be obtained using the gas pump-out method?
8. What does the desorption cooling method mean?
9. Follow the analogy between the processes of desorption cooling and isentropic expansion.

7

LOW TEMPERATURE THERMODYNAMIC CYCLES

Thermodynamic cycles of cryogenic installations are numerous and various. Ideal cycles are examples for comparison and improving of real systems.

7.1. Characteristics of thermodynamic cycles

The degree of perfection of refrigerated and cryogenic installations, as well as of cycles to be realized in them, is distinguished by the following parameters.

1) Specific refrigerating capacity or power of a cycle ($J{\cdot}kg^{-1}$) is

$$q_2 = \frac{Q_2}{G}, \tag{7.1}$$

where Q_2 is the heat removed per unit time, W; G is the mass flow-rate of working substance circulating in a system, $kg{\cdot}s^{-1}$.

2) The performance energy ratio is

$$\varepsilon = \frac{Q_2}{L} = \frac{q_2}{l}, \tag{7.2}$$

where L is the spent energy, W; l is the spent energy per unit mass of circulating substance, $J{\cdot}kg^{-1}$.

For the Carnot ideal refrigeration cycle to provide heat transfer from a lower temperature level T_c to an upper level T_0, we have

$$\varepsilon_{id} = \frac{q_2}{l_{min}} = \frac{T_c}{T_0 - T_c}. \tag{7.3}$$

3) A liquefaction coefficient characterizes liquefaction cycles:

$$y = \frac{G_{liq}}{G}, \tag{7.4}$$

where G_{liq} is the liquid mass flow-rate, $kg \cdot s^{-1}$; G is the mass flow-rate of the substance to enter the installation; y may be interpreted as the ratio of the cycle cold (minus losses) to the minimum cold required for gas liquefaction.

4) Specific energy consumption l_0 is the ratio of the spent energy l to the obtained effect (amount of cold q_2 in refrigeration cycles, liquefaction coefficient y in liquefaction cycles, mass fraction M_i of separated product in gas separation cycles).

5) Thermodynamic efficiency characterizes the effectiveness of a real cycle, as compared to an ideal one:

$$\eta_T = \frac{l_{min}}{l} = \frac{\varepsilon}{\varepsilon_{id}}. \tag{7.5}$$

By an ideal cycle is understood a reverse cycle which is thermodynamically most ideal for solving this specific problem. The Carnot reverse refrigeration cycle (Fig. 7.1a) is perfect for thermostating at given values of temperatures T_c and T_0, which the cycle in Fig. 7.1b, is ideal for gas liquefaction.

For ideal liquefaction cycle, the gas is isothermally compressed in a compressor (process 1-2) with work

$$l_{comp} = T_0 (s_1 - s_2) - (h_1 - h_2) \tag{7.6}$$

and is adiabatically expanded in an expander (process 2-3) with work

$$l_{exp} = h_2 - h_{l3}, \tag{7.7}$$

where s is specific entropy; h is specific enthalpy; h_{l3} is

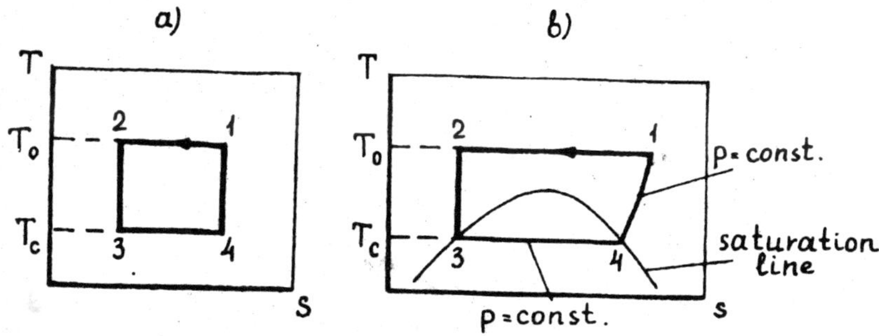

Fig. 7.1. Ideal thermostating (a) and liquefaction (b) cycles.

liquid enthalpy on the saturation line at temperature T_c.

The cooling process might be set up more naturally, i.e. by cooling at $p = const = p_1$ (process 1-4, Fig. 7.1b) and by condensation at $p = const = p_1$ (process 4-3) with heat removal (per unit mass)

$$q_2 = h_1 - h_{13} . \qquad (7.8)$$

Thus, the ideal reversible liquefaction cycle 1-2-3-4-1 with a performance energy ratio is

$$\varepsilon_{id} = \frac{q_2}{l_{min}} = \frac{q_2}{l_{comp} - l_{exp}} = \frac{h_1 - h_{13}}{T(s_1 - s_2) - (h_1 - h_{13})} . \qquad (7.9)$$

Over a broad temperature range $(T_o - T_c)$, the cascade cooling method is used. Over the temperature range $(T_o - T_c)$ not one but several reverse cycles proceed, each of which removes heat from the bottom (by temperature) to the top cycle. Very low temperatures may thus be attained, down to the liquid helium value.

Let us consider for e.g. cascade liquefaction. In the first cycle (from ambient temperature T_0), ammonia or Freon usually serves as the working substance. In the second cycle ethylene is condensed under pressure in an ammonia or

Freon evaporator. Ethylene evaporates at 173 K and liquefies compressed methane serving as the third-cycle cryoagent. Methane, to evaporate at 112 K, liquefies compressed nitrogen in the fourth cycle. The temperature may be depressed to 63 K by vacuum evaporation of nitrogen. As this temperature lies above the critical value for hydrogen (33 K), hydrogen liquefaction by a simple continuation of cascade cycle is not possible. Hydrogen compressed to a pressure of 15 MPa is cooled to 63 K by evaporating nitrogen and reverse hydrogen flow and is then throttled with forming a vaporliquid mixture (fifth cycle). The temperature 14 K (vacuum evaporation of hydrogen) is above the critical value for helium (5 K). Therefore, in the sixth cycle, helium compressed to 4 MPa is first cooled by evaporating hydrogen, that is reverse helium flow, and then throttled. As a result, it is partially liquefied. A temperature of 0.5 K may be attained by vacuum evaporation of helium.

In principle, cascade liquefaction is one of the most economical methods. The following values of specific energy consumption (kW·h per 1 kg of liquid nitrogen) have been calculated in nitrogen liquefaction by different methods (no regard to heat fluxes due to imperfect insulation and hydraulic pressure loss):

Cascade method NH_3 - C_2H_4 - CH_4 - N_2 . . . . . . 0.54
Double nitrogen throttling and preliminary
ammonia cooling cycle 0.85
Isoentropic nitrogen expansion cycle 0.88

Consider, in more detail, the commonly used thermodynamic cycles of refrigerated and cryogenic installations.

7.2. Throttling Cycles

7.2.1. Single throttling cycle (Linde cycle)

A schematic of the single throttling cycle and its circuit are shown in Fig. 7.2 (temperature - entropy diagram). Figures on the circuit correspond to working substance states

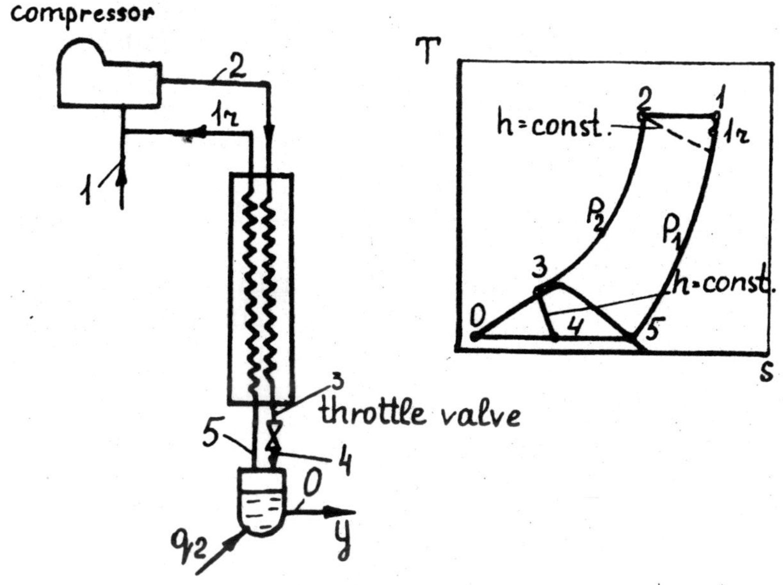

Fig. 7.2. Schematic of a single-throttling cycle.

on the **T-s** diagram.

The compressed gas is cooled by the reverse gas flow in a heat exchanger, then expanded in a throttle valve and some amount is liquefied. The cycle energy balance per unit compressed gas is:

$$q_{sup} = 1 \times h_2 + q_a$$

is supplied and

$$q_{rem} = y\, h_0 + (1\text{-}y)\, h_1$$

is removed.

Hence

$$y = \frac{h_{1r} - h_2 - q_a}{h_{1r} - h_0}\ ; \qquad q_2 = h_{1r} - h_2 - q_a, \qquad (7.10)$$

where q_2 is the specific refrigerating capacity of the cycle; h_{1r}, h_2, h_0 and h_1 are the gas enthalpies at the points indicated (Fig. 7.2); y is the portion of liquid per unit throttled gas; q_a is the cycle heat from the ambient medium due to imperfect insulation (so-called cold losses via insulation to the ambient medium).

For an ideal cycle, $q_a = 0$ and $h_{1r} = h_1$. For a real cycle $q_a > 0$ and $h_{1r} < h_1$.

From (7.10) it is seen that the refrigerating capacity of a cycle depends solely on the enthalpy difference of expanded and compressed gases at the temperature of a "hot" end of a recuperative heat exchanger, i.e. on the isothermal throttling effect.

Frequently, this cycle is used not to produce liquid cryoagent as product but just to cool a device by removing heat q_2 from it (Fig. 7.2, dashed line). In this case, the cycle operates under so-called refrigeration conditions. In contrast to gas liquefaction, under these conditions liquid is not removed ($y = 0$) and the backward flow of the gas equals the forward flow.

..2.2. Single-throttling and preliminary cooling cycle

A single-throttling and preliminary cooling cycle is shown in Fig. 7.3. Unlike the described above cycle, this cycle uses extra cryoagent (e.g. ammonia) to produce preliminary cooling for the main cycle. For this cycle we have

$$y = \frac{h_{7r} - h_3 - q_a}{h_{7r} - h_0} . \tag{7.11}$$

For ideal conditions, $q_a = 0$ and $h_{7r} = h_7$.

The refrigerating cycle capacity is determined by the isothermal throttling effect at a cryoagent temperature after the letter has been cooled by a foreign cryoagent.

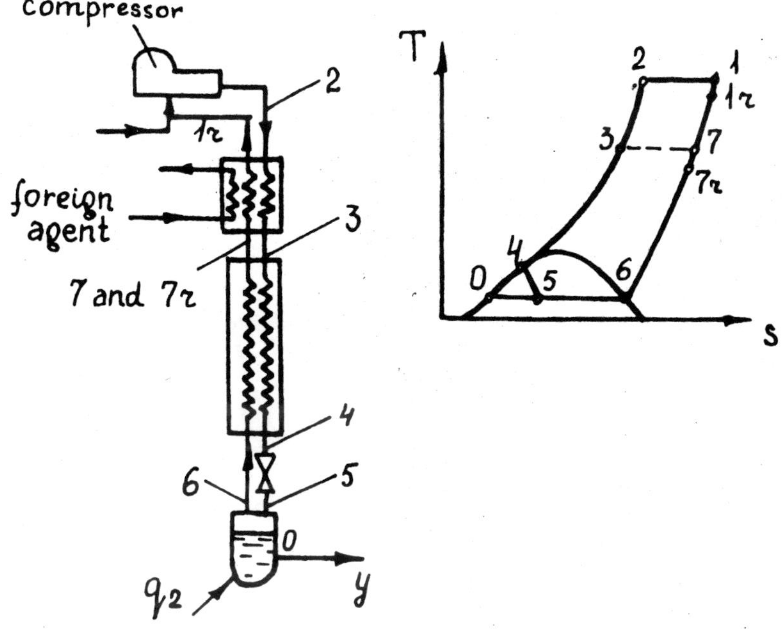

Fig. 7.3. Schematic of a single throttling and preliminary cooling cycle.

The amount of heat transferred to a foreign cryo-agent is

$$\mathbf{q_f = h_2 - h_3 - (1 - y)(h_{1r} - h_{7r})} . \tag{7.12}$$

For a refrigerator $(\mathbf{y = 0})$ we have

$$\mathbf{q_f = h_2 - h_3 - (h_{1r} - h_{7r})} . \tag{7.13}$$

Preliminary cooling is essential for liquefaction of gases whose inversion temperature is below ambient in the cycles analyzed above. So, these gases must be cooled below 180 K and 40 K, respectively, to liquefy hydrogen and helium.

7.2.3. Double-throttling cycle

A double-throttling cycle (with circulation at intermediate pressure) is shown in Fig. 7.4. The gas is

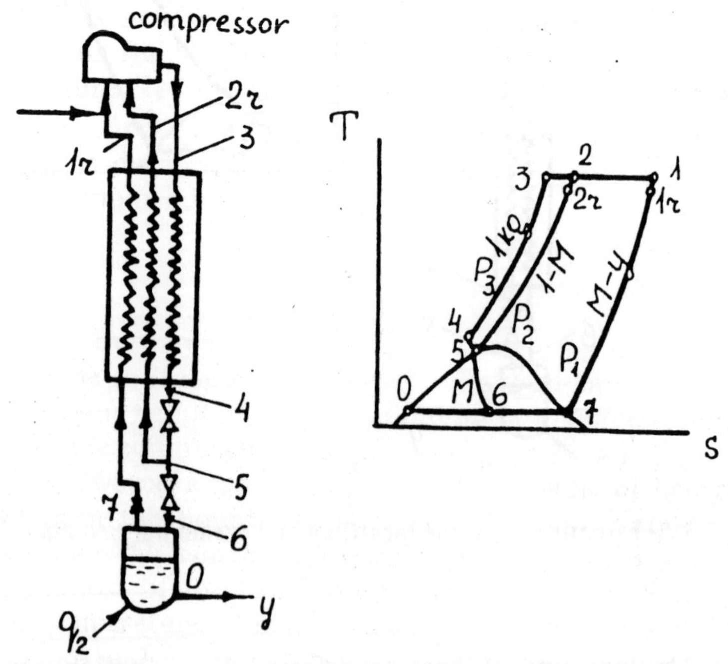

Fig. 7.4. Schematic of a double-throttling cycle.

compressed to pressure p_2 and, after circulating flow is added, it is compressed to p_3. After cooling, in a heat exchanger (line 3-4) the gas is throttled to an intermediate pressure p_2. Some quantity of gas **(1 – M)** passes via the heat exchanger in the opposite direction (line 5-2) and enters the compressor. The rest of the gas **(M)** is throttled to p_1 (usually 0.1 MPa). In this case, it is partially liquefied (liquid fraction **y**).

For a cycle

$$y = \frac{h_{2r} - h_3 + M(h_{1r} - h_2) - q_a}{h_{1r} - h_0};$$

$$q_2 = h_{2r} - h_3 + M(h_{1r} - h_2) - q_a \,. \quad (7.14)$$

Under ideal conditions $q_a = 0$ and $h_{1r} = h_1$; $h_{2r} = h_2$.

This cycle may use a preliminarily cooled foreign cryoagent. Here, formules (7.14) are supplemented with isothermal throttling effects $(h_{2r} - h_3)$ and $(h_{1r} - h_2)$ at the final temperature of preliminary cooling.

In a preliminary cooling cycle, M cannot be very small, otherwise heat transfer is not possible at the cold end of the recuperative heat exchanger.

7.2.4. Throttling multicomponent-gas mixture cycle.

Application of mixtures in throttling systems is, to a certain extent, connected with successes in liquefaction and separation of natural gases. Recently, throttling cycle are being employed in micro-cryogenic systems to cool radio-electronic units. Mixtures containing a highly evaporating component with large isothermal throttling cycle offer an increase in total throttling effect. Of most use are mixtures with low-evaporating component, e.g. nitrogen or argon with added saturated hydrocarbons (CH_4, C_2H_6, C_3H_8) or else with refrigerants - 14, 13, as well as with krypton and xenon. Such mixtures provide nitrogen temperatures to be obtained in the refrigerating system evaporator. Correct choice of mixture gives component condensation in forward flow and their evaporation in backward flow in a heat exchanger over a wide temperature range. This yields an optimum temperature head along the entire heat exchanger and, hence, low thermodynamic loss under heat transfer and throttling condition.

Of significance is the avoidance of elevated temperature of an evaporating working mixture in the evaporator through the addition of a high-evaporating component. A small amount of low-evaporating component (neon, hydrogen, helium) is added to a mixture to depress the evaporating temperature. After throttling, the partial pressure of the remaining components is decreased due to small solubility of the low-evaporating component in liquid, thus resulting in reduced evaporating temperature. Thus, temperatures $77 \div 80$ K and below ($65 \div 70$ K) are attained by using nitrogen-base mixtures, but with somewhat elevated evaporation temperature in the flow via the evaporator.

It is found that some nitrogen-hydrocarbon mixtures at the nitrogen temperature are subdivided into two liquid phases: one is almost pure nitrogen while the other is low nitrogen-content hydrocarbon. Also, eutectic alloys are formed in a solid body-liquid system. This yields a very low temperature in the evaporator with no solid phase separation.

Figure 7.5 interprets the refrigerator processes, with a mixture forming two liquid phases. A closed stripe (Fig. 7.5b) is produced according to the schematic in Fig. 7.5a and displays a gas-to-liquid phase ratio as a process proceeds. After throttling, the liquid phase L_1 composed mainly of a low-evaporating component (nitrogen) is almost completely evaporated at a constant temperature ($T_5 = T_4$). Another liquid phase L_2 composed of highly evaporating components passes via the evaporator practically, unchanged and is evaporated only in the heat exchanger. Thus, high evaporating components providing a large total throttling effect of a mixture and a required heat equivalent ratio in the heat exchanger barely affect the evaporator process. The thermodynamic efficiency of throttle refrigerators operating with mixtures over the temperature range $75 \div 80$ K is several times higher than for pure substance (nitrogen).

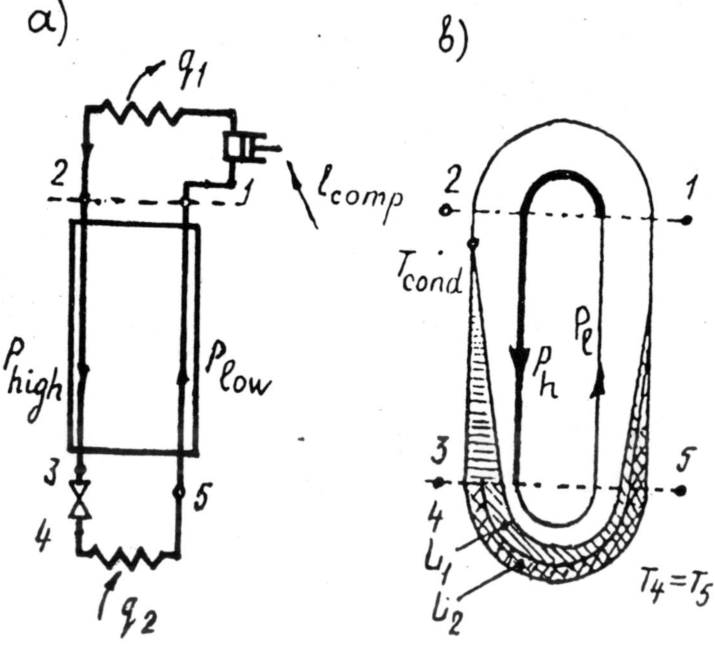

Fig. 7.5. Schematic of a throttle refrigerator using mixtures to produce separating liquid phases.

7.3. Expansion and Work Cycles

7.3.1. Gas refrigerator cycles

Gas refrigerator cycles are shown in Fig. 7.6. A gas compressed to p_2 is cooled in a refrigerator (process 2-3) and is then expanded adiabatically in the expander (3-4) to p_1. Cold gas is directed to the cooling chamber, and heated (4-1), whereupon it is adiabatically compressed (1-2).

The performance energy ratio is

$$\varepsilon = \frac{q_2}{l} = \frac{h_1 - h_4}{(h_2 - h_1) - (h_3 - h_4)} = \frac{T_1 - T_4}{(T_2 - T_1) - (T_3 - T_4)}; \quad (7.15)$$

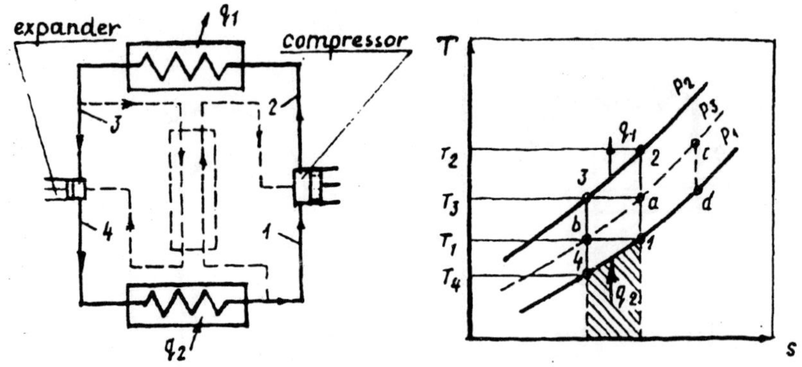

Fig. 7.6. Schematic and prosesses of a gas refrigerating machine (b-a-c-d-1-4 regenerative cycle).

A gas refrigeration cycle has a substantial drawback: refrigerating capacity $\mathbf{q_2}$ is realized at a variable temperature ($\mathbf{T_4}$ to $\mathbf{T_1}$). Since gas cooling to point 3 (Fig. 7.6) requires refrigerant with $\mathbf{T_3}$, and cooling must procees at a temperature not higher than $\mathbf{T_1}$, the appropriate Carnot cycle would have a performance evergy ratio $\varepsilon = \mathbf{T_1 / (T_3 - T_1)}$, i.e. substantial. To reduce gas compression and to facilitate operation, as well as to decrease temperature $\mathbf{T_4}$, regenerators (heat exchangers) mounted ahead of the expander are used. Here, the forward flow to enter an expander is cooled by the backward flow leaving the cooling chamber and directing to the compressor. Flows in the so-called regenerative cycle are shown by the dashed lines in Fig. 7.6. Energy consumption in a regenerative cycle is the same as in an ordinary cycle.

7.3.2. Reverse Stirling cycle

Reverse Stirling cycle - gas cryomachine cycle. The schematic of a cryomachine and the **p-V** diagram for a cycle are shown in Fig. 7.7.

A regenerator **R** (heat-accumulating mass) is located

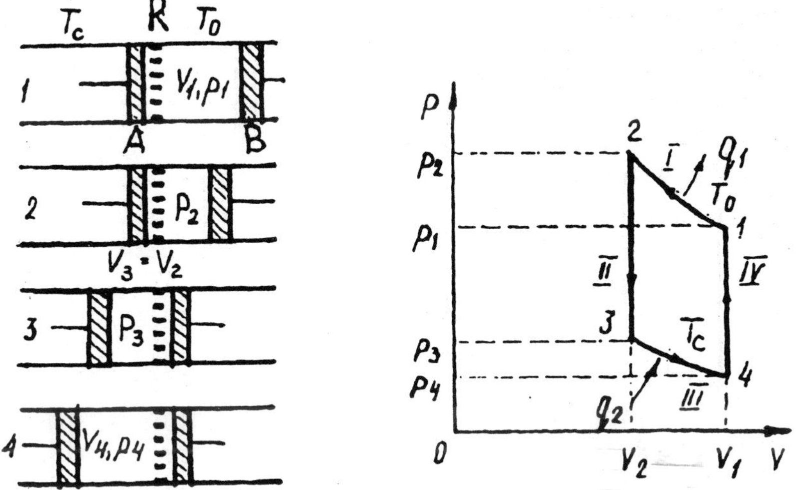

Fig. 7.7. Reverse Stirling cycle.

between pistons **B** (the so-called hot side of the cryomachine) and **A** (the so-called cold side). Gas going from the hot to the cold volume is cooled in the regenerator; backward flow gas is heated in the regenerator. In the cycle the following processes occur:

I, isothermal compression 1-2 (heat q_1 is removed by cooling water);

II, isochoric cooling 2-3 (via the regenerator);

III, isothermal expansion 3-1 (heat q_2 is removed from a cooled body);

IV, isochoric heating 4-1.

For an ideal gas, the cycle is equivalent to the Carnot cycle between isotherms T_0 and T_c. At isothermal processes work is equivalent to heat. The specific refrigerating capacity q_2 is equal to the isotermal expansion work l_{exp}.

$$q_2 = l_{exp} = R T_2 \ln \frac{v_1}{v_2} .$$

The isotermal compression work is equal to the heat, removed from gas:

$$l_{comp} = q_1 = RT_0 \ln\frac{v_1}{v_2} .$$

The performance energy ratio is

$$\varepsilon = \frac{q_2}{q_1 - q_2} = \frac{T_2}{T_0 - T_2} . \quad (7.16)$$

Usually, helium or hydrogen with a degree of expansion 2 ÷ 2.5 (e.g. from 3.5 to 1.6 MPa) is used as the working substance.

The Stirling cryomachine generates liquid air, condensed on the outer finned side of the cylinder of the cold volume.

The firm Philips (Holland), the first to manufacture such cryomachines, is now producing various designs of cryomachines with various refrigerating capacities of 2100 $kJ{\cdot}h^{-1}$ (5 litres of liquid air per hour), 3300, 12500, and 83000 $kJ{\cdot}h^{-1}$ at 77 K. Energy consumption is 1.1 kW·h per 1 kg of liquid air.

Two-stage cryomachines are designed to attain lower temperatures. Fig. 7.8 shows the schematic of a cryomachine of this type. As in the ordinary one-stage cryomachine, test gas is compressed in the hot volume V_h and, after passing via water refrigerator (1) it enters first regenerator (2) and then at a temperature T_{int} enters an intermediate volume V_{int} where about 1/5 of the gas is expanded. Here, some refrigerating capacity q_{int} may be removed at a temperature $T_{int} = 50 \div 80$ K using heat exchanger (3). The remaining gas passes via the second regenerator (4) and is expanded in the low-temperature volume V_c where a refrigerating capacity q_c is removed at the lower temperatrure T_c (down to 15 K) using heat exchanger (5). Cold produces at T_{int} compensates all losses of the first regenerator and offers effective gas cooling in second regenerator (4).

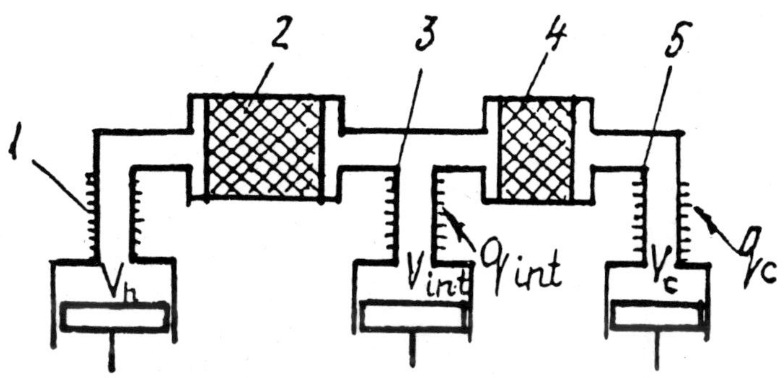

Fig. 7.8. Schematic of a two-stage cryogenic gas machine operating according to modified Stirling cycle: 1, water refrigerator; 2,4, regenerators; 3,5, heat exchangers.

In addition, the possibility of producing cold at two temperatures is especially profitable for gas liquefaction since it allows preliminary cooling of a gas before it is condensed, thus reducing thermodynamic heat transfer losses.

Cryogenic machines operating according to the Stirling cycle are most profitable and effective for recondensation of saturated vapor formed during evaporation of liquid cryogenic products. In this case, a cryomachine power is used with minimal unreversibility; heat is transferred at constant temperature and minimum possible temperature head.

7.4. Combined Expansion and Throttling Cycles

7.4.1. Average pressure cycle (Claude cycle) and high pressure cycle (Heyland cycle)

A schematic of an installation in which expansion and throttling processes occur is shown in Fig. 7.9 in **T-s**

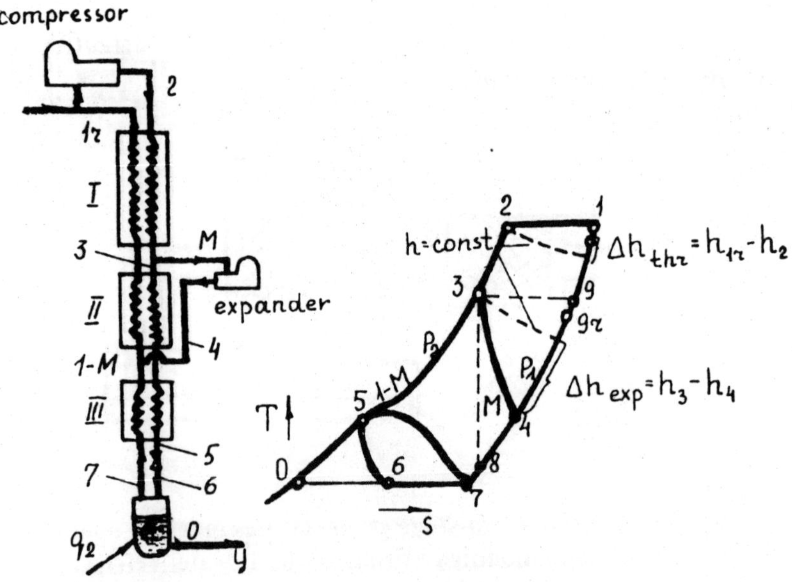

Fig. 7.9. Schematic of an expander device and process on a T-s diagram.

coordinates.

An average-pressure (1.5 ÷ 6.0 MPa) compressed gas after being cooled in heat exchanger (I) is subdivided into two parts. One **(M)** enters the expander to be expanded with external work and the rest **(1 – M)** is cooled in heat exchangers (II) and (III) and is then expanded in the throttle valve. On the **T-s** diagram, line (3-4) shows a real process in the expander. The dashed line (vertical 3-8) corresponds to an ideal process in the expander (**s = const**).

The energy balance per unit compressed gas is:

$$\mathbf{q_{sup} = 1 \times h_2 + M\,h_4 + q_a}$$

is supplied and

$$q_{rem} = y h_0 + (1 - y) h_{1r} + M h_3$$

is removed.

Hence, the liquid amounts to

$$y = \frac{h_{1r} - h_2 + M(h_3 - h_4) - q_a}{h_{1r} - h_0}; \tag{7.17}$$

and the refrigerating capacity is

$$q_2 = h_{1r} - h_2 + M(h_3 - h_4) - q_a = \Delta h_{thr} + M \Delta h_{exp} - q_a .$$

The adiabatic efficiency of the expander η_s to characterize deviation of a real expansion process from ideal **(s = const)** is

$$\eta_s = \frac{h_3 - h_4}{h_3 - h_8}. \tag{7.18}$$

The η_s values are:

- air piston expanders . . . . 0.6 ÷ 0.75;
- helium expanders . . . . 0.75 ÷ 0.9;
- air turbo-expanders (Kapitza) . . . 0.8 ÷ 0.85.

The optimum gas temperature ahead of the expander depends on gas pressure. From eqn. (7.17) it is seen that a maximum amount of a gas (part **M**) may be directed to the expander to increase a cycle's refrigerating capacity. However, in order that the backward flow can cool the forward flow of **(1 – M)**, the temperature difference at the hot heat exchanger end must not exceed normal value (5 K).

Each forward-flow pressure is corresponded with gas optimum part **M** directed to the expander as well as with its optimum temperature before the expander when the liquid fraction is maximum and specific flow-rate minimum.

Also, an expansion cycle is used where gas is compressed to a pressure of 20 ÷ 16 MPa. In this cycle (Fig. 7.9), some gas may enter the expander either after compression, (in this case heat exchanger I is absent, temperature at point 3 will be equal to that after compression, i.e. amounting to about 300 K) or after preliminary cooling by 50 ÷ 70 K.

This cycle is called the high-pressure expansion cycle (Heyland cycle) and is widely used to produce liquid oxygen from air. Such a cycle (and its modifications) is one of the most energy profitable for gas liquefaction. Equation (7.17) is valid for predicting a high-pressure cycle.

Usually, gas expansion in a cylinder of piston expanders ceases not at 0.1 MPa but at a higher pressure (0.5 ÷ 1.0 MPa). Further expansion proceeds by throttling in discharge valves. Gas expansion up to 0.1 MPa in an expander is associated with sharply increasing cryomachine overall dimensions and other difficulties. Recently, high-pressure installations are adopting turbo-expanders.

7.4.2. Low-pressure cycle with expansion in turbo-expander (Kapitza cycle)

Low-pressure cycle with expansion in turbo-expander is shown in Fig. 7.10. Gas (air) is compressed to 0.6÷0.8 MPa (process 1-2) and then enters regenerators where it is cooled (2-3). The air is then subdivided into two parts: most air is to be expanded in the expander (3-4r), while the remainder is sent to the heat exchanger-condenser where this quantity of air condenses (3-5). The liquid is then throttled (5-6) to 0.1 MPa. After the turbulizer, the gas flows in the opposite direction to the heat exchanger-condenser and then to the regenerators (4r-1r). On the **T-s** diagram, line (3-4r) corresponds to a real process in the expander, and line (3-4) to the process at **s = const**.

On throttling, the liquid fraction is

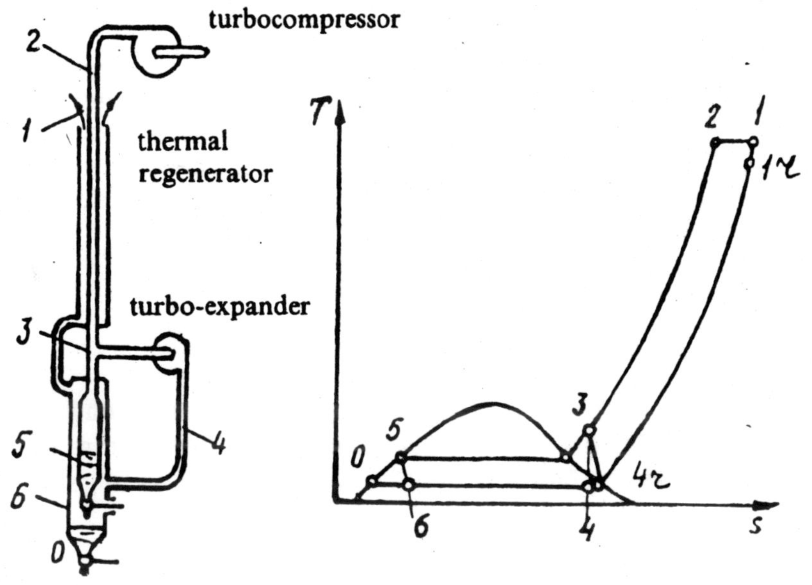

Fig. 7.10. Schematic of a low-pressure expander device.

$$y = \frac{(h_{1r} - h_2) + (1 + a\,y)(h_3 - h_4) - q_a}{h_{1r} - h_0}, \qquad (7.19)$$

where **a** is a coefficient allowing for partial liquid evaporation during throttling from condensation pressure to atmospheric that.

For ideal cycle conditions, $q_a = 0$ and $h_{1r} = h_1$.

A turbo-expander designed by P. Kapitza has an efficiency of 0.8 or even more. For one turbine wheel, permissible inlet air pressure is $0.6 \div 0.8$ MPa. An air one-stage expander is most profitable at this pressure.

From calculations, the energy consumption in expander-equipped installations, operating efficiency of 0.8, is about 1.6 kW·h per 1 kg of liquid air at an isothermal compressor efficiency of 0.6.

Use of low-pressure air alone in a cycle offers regenerators which are characterized by low sub-recuperation and requires no preliminary aircleaning from CO_2 and H_2O. Application of turbo-machines alone means that high-efficiency units can be designed. Use of high-efficiency turbo-compressors reduces the energy consumption for gas liquefaction.

7.4.3. Cascade expansion cycles

Using several expansion processes proceeding at different temperatures, a cryogenic gas liquefaction cycle with smaller reversibility may be performed. Fig. 7.11 shows

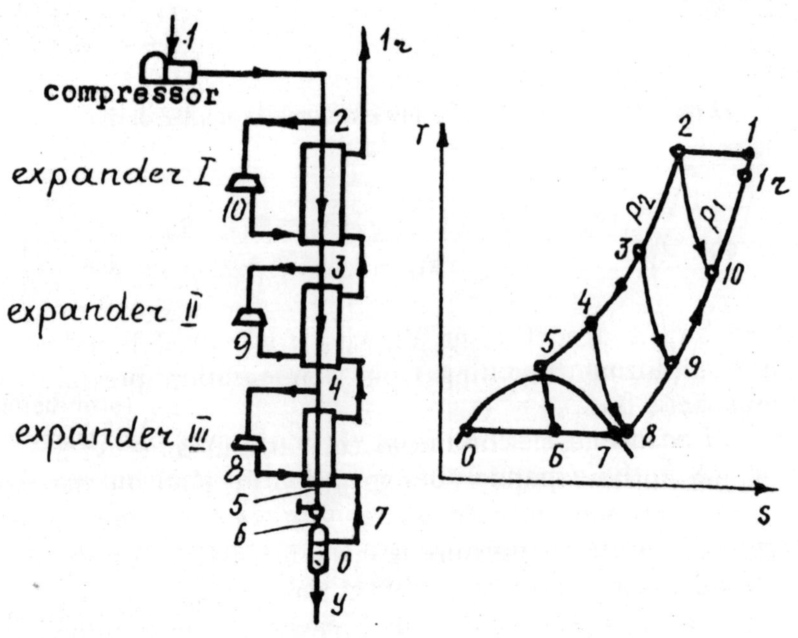

Fig. 7.11. Schematic of a cascade expander.

the schematic of a three-expander device. Processes are shown in **T-s** coordinates. For test bodies not strongly different from ideal gases, it may be considered to first approximation that approximately the same amount of gas enters each expander. The number of expanders in a cascade depends on the degree of gas expansion, the number of expanders decreasing with increasing degree of expansion.

A version is possible where the installation is equipped with a cascade of expanders. In such a system, some amount of gas to be throttled is compressed to a higher pressure, compared to that expanding in expanders. Such installations have found use only in helium liquefaction.

Cryogenic cycles where flow throttling is replaced by expansion in an expander, or in an ejector are employed. Asengineering problems on two-phase expanders have been solved, such cycles are also applied for helium liquefaction in refrigeration conditions. In this case, the refrigerating capacity is increased by as many as $1.5 \div 2$ times, while helium liquefaction undergoes a 1.3-fold enhancement.

Questions

1. What parameters characterize the efficiency of low-temperature cycles?

2. Derive the formula for the coefficient of per-formance of a refrigerating machine for an ideal liquefaction cycle.

3. What is the set-up of the cascade cooling principle?

4. Follow the Linde cycle on the **T-s** diagram and derive the formulas for the liquefaction coefficient and specific refrigerating capacity of a cycle.

5. What is the role of preliminary cooling in a throttling cycle?

6. What are the specific features of cycles using multicomponent gas mixtures?

7. What are the specific features of gas refrigeration machines?

8. How does the Stirling gas machine operate?

9. Derive the formulas for the liquefaction coefficient and specific refrigerating capacity of a combined expanding and throttling cycle.

10. What are the specific features of Kapitza's turbo-expander?

8

ELECTROMAGNETIC METHODS OF COOLING

Specific methods of cooling are associated with processes in electric and magnetic fields.

Consider following electromagnetic methods:

- thermoelectrical cooling;
- galvanothermomagnetic cooling;
- adiabatic demagnetization method;
- adiabatic magnetization of a superconductor;
- nuclear demagnetization.

8.1. Thermoelectrical cooling

This method is based on the Peltier effect; on conduction of a current in a circuit composed of two different conductors, heat **Q** is released or absorbed at the contact points:

$$\mathbf{Q} = \alpha \mathbf{J} \mathbf{T},$$

where **J** is current, **T** is the junction temperature, α is the thermal EMF coefficient dependent on type of materials.

The Peltier effect is the opposite of the Seebeck effect used in thermocouples. Physically, this phenomenon implies that the energy needed by free electrons to participate in the electric current in different conductors depends on temperature.

If there is temperature drop, then electrons have greater energy at the "hot" end than at the "cold" end; as a result, electrons move to the "cold" end where charge is stored and a back electromotive force develops. The difference of these electromotive forces in different conductors is the cause of thermal EMF.

If in a circuit the direction of electrons is such that lower-energy electrons having removed heat from the ambient atoms transport it to another part of the circuit, then one junction is cooled and another is heated (Peltier effect). In this way a pair of conductors like this may be used as a refrigeration unit in which electron gas transporting the energy from a cold to a hot junction serves as the working fluid. This effect is not large for ordinary thermocouples. Academician A. Ioffe and co-workers showed that use of semiconductor refrigerants may yield a more than ten-fold high cooling efficiency.

As its main structure the thermocooling device uses, a thermoelement (Fig. 8.1) composed of two semiconductor branches "a" and "b" in series, one of which possesses electron (n) conduction, the other hole (p) conduction.

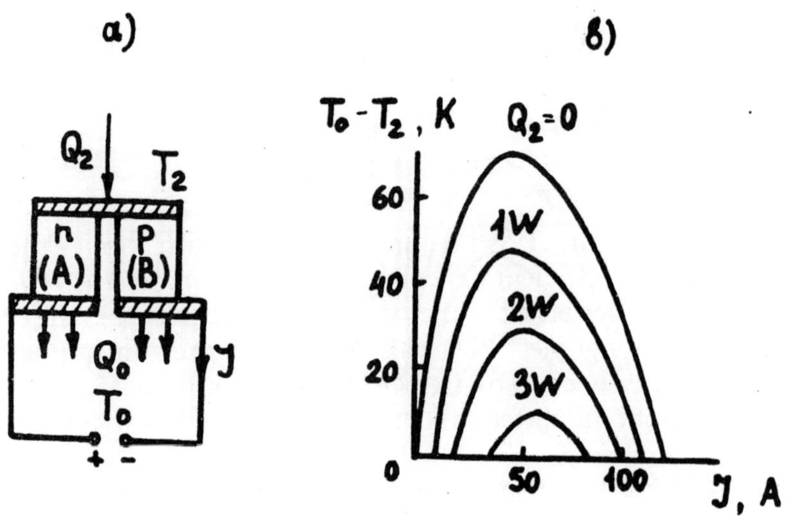

Fig. 8.1. Schematic of elementary Peltier thermocoolant (a) and perfomance of a thermal battery at T_0=289 K, Z=0.00267 K^{-1} (b).

In conducting current **J**, one junction absorbs Peltier heat Q_2 at T_2, the other releases heat Q_0 removed at T_0. As mentioned,

$$Q_2 = (\alpha_a - \alpha_b)\, J\, T_2 . \tag{8.1}$$

In the current conductor Joule heat is released, thus reducing Q_2. It is found that to a first approximation at each junction half the entire Joule heat is released, i.e.

$$Q_j = 0.5\, J^2 R , \tag{8.2}$$

where **R** is the total electric resistance of the the branches.

The amount of heat absorbed on the cold junction is also decreased due to heat flux Q_{heat} from the hot junction via the conductor branches

$$Q_{heat} = K\,(T_0 - T_2) , \tag{8.3}$$

where **K** is the total heat conduction of the branches.

Fig. 8.1 shows a plot of $(T_0 - T_2)$ versus **J** at different Q_2.

The performance energy ratio of a thermal battery is

$$\varepsilon = \frac{Q_2}{L} = \frac{(\alpha_a - \alpha_b) T_2 J - 0.5 J^2 R - K(T_0 - T_2)}{J^2 R + (\alpha_a - \alpha_b)(T_0 - T_2) J} , \tag{8.4}$$

where the denominator is the power spent for the Joule heat and for overcoming the thermal EMF.

A maximum value ε_{max} and appropriate optimum current

$$\varepsilon_{max} = \frac{T_0 (M - T_0 / T_2)}{(T_0 - T_2)(M + 1)} ; \tag{8.5}$$

$$J_{opt} = \frac{(\alpha_a - \alpha_b)(T_0 - T_2)}{R(M - 1)}, \qquad (8.6)$$

where

$$M = \sqrt{1 + Z(T_0 + T_2)/2};$$

$$Z = \frac{(\alpha_a - \alpha_b)^2}{R\,K} = \left(\frac{\alpha_a - \alpha_b}{\sqrt{\lambda_a \rho_a} + \sqrt{\lambda_b \rho_b}}\right)^2. \qquad (8.7)$$

If the branches of the thermo-element are assumed to have the same geometrical sizes, the same heat conduction ($\lambda_a = \lambda_b = \lambda$), the same specific electric resistance ($\rho_a = \rho_b = \rho$), and the thermal EMF coefficients are assumed to be equal and opposite in sign ($\alpha_a = -\alpha_b$), then under these conditions we have a criterion of the thermo-electrical cooling efficiency for the thermo-element

$$Z = \frac{\alpha^2}{\lambda\rho} = \frac{\alpha^2 \sigma}{\lambda}, \qquad (8.8)$$

where $\sigma = 1/\rho$ is electric conductivity.

Since a single thermo-element produces only a low amount of refrigerating power, thermal batteries composed of thermoelements (hundreds or thousands in number) in series are used to obtain rather large Q_2.

The magnitude Z relating thermo-electric, electric, and thermal properties of a substance is the thermo-electrical cooling efficiency criterion. Thus, it is significant to find materials with high Z. The relation σ/λ for all metals is approximately the same and is smaller in semiconductors than for metals. However, their α is much greater and, in this way, Z is also larger. This fact has allowed semiconductors to be successfully employed in thermal cooling. At present, solid solutions Bi_2Te_3 - Bi_2Sb_3 and Bi_2Te_3 - Sb_3Te_3 are considered to be the best materials for thermo-elements. For them, $Z \sim T^{0.5}$ for low temperatures ($T < 250$ K) and $Z \sim T^{-0.5}$ for relatively high values ($T \geq 250$ K). For thermo-elements of various semiconductors Z is $1.2\times10^{-3} \div 4\times10^{3}$ 1/K at moderate temperatures. The magnitude Z should be taken at the mean temperature.

Cascade cooling is used to increase a temperature drop and refrigerating unit efficiency. Several thermal batteries are used to cool each other in succession.

The total temperature drop attained by refrigerating units is usually $(T_0 - T_2) < 100$ K (at $T_2 > 150$ K).

As an example, it is possible to give characteristics of a thermal refrigerating unit having a maximum performance energy ratio ε_{max} at $Z = 1.93\times10^{-3}$ K and $T_2 = 273$ K (Table 8.1).

Table 8.1. Optimal characteristics of refrigerating unit at different temperature drops.

$(T_0 - T_2)$, K	ε_{max}	$\eta_T = \varepsilon_{max} / \varepsilon_{id}$, %
5	5.33	9.75
20	1.04	7.65
40	0.32	4.65
60	0.08	1.75

To obtain cryotemperatures, thermocooling elements may be used as the last cooling stage combined, for example with the Stirling cycle cryomachine. It is interesting to note that for a bismuth-antimony alloy $Z = 4.8\times10^{-3}$ K^{-1} at 77 K while at 300 K, $Z = 1.8 \times 10^{-3}$ K^{-1}. An alloy may be effectively used over a cryogenic temperature range.

8.2. Galvanothermomagnetic cooling (Ettingshausen effect)

In 1886, Nernst and Ettingshausen discovered this effect: in a conductor placed in a magnetic field and having a temperature gradient perpendicular to this field, an electric field (Nernst effect) develops in the third mutually perpendicular direction. Vice versa, if in a conductor the current is perpendicular to a magnetic field, then a temperature gradient appears in the third mutually perpendicular direction (Ettingshausen effect) (Fig. 8.2).

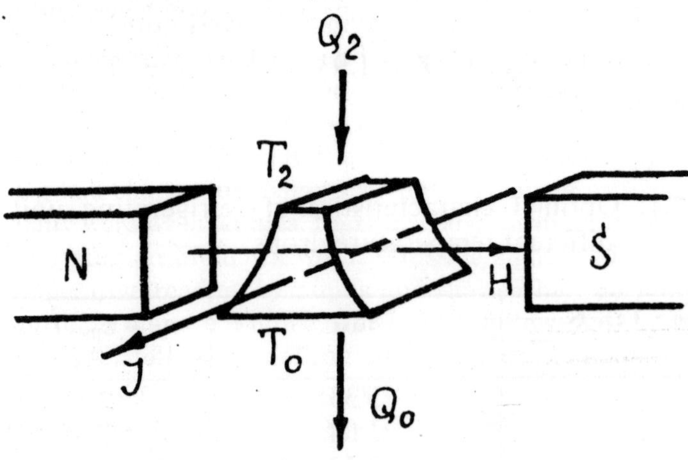

Fig. 8.2. Illustration of the Ettingshausen effect.

The Nernst (**N**) and Ettingshausen (**P**) coefficients are equal to

$$\mathbf{N} = \frac{\mathbf{E}}{\mathbf{H}\,\Delta\mathbf{T}} \; ; \qquad \mathbf{P} = \frac{\Delta\mathbf{T}}{\mathbf{N}\,\mathbf{J}} \, . \qquad (8.9)$$

Here $\mathbf{E}$ is the electric field strength, $\mathbf{H}$ is the magnetic field intensity, $\mathbf{J}$ is the current, $\Delta T = T_0 - T_2$ is the temperature gradient.

Since a thermal refrigerating unit is composed of one electrical branch, it is quite simple to design a cascade refrigerator. The shape of such a refrigerator with an infinite number of stages is shown in Fig. 8.2: upper area, "cold" side (Q_2, T_2); lower area, "hot" side (Q_0, T_0). Such a conductor shape is stipulated by the heat flux which, moving from the cold surface, is increased by the work spent on heat conversion.

It may be shown that the Ettingshausen cooling effect is calculated by the relations set for the Peltier effect, of course replacing the appropriate separate terms. So, the criterion

$$Z = N H^2 \sigma / \lambda , \qquad (8.10)$$

is included, similar to (8.8).

Maximum ΔT is

$$\Delta T_{max} = 0.5 Z T_0^2 . \qquad (8.11)$$

An alloy of 97% Bi and 3% Sb is used as conductor material. For monocrystals made of such an alloy, the values of Z are smoothly increased from $0.2 \times 10^{-3}\ K^{-1}$ at 300 K to maximum $3.5 \times 10^{-3}\ K^{-1}$ at 100 K in a magnetic field of 1.0 T.

Calculation shows that in a 1.0 T field, temperature reduction $\Delta T = 9 \div 12$ K may be attained at $T_0 = 77$ K on the hot side. A system based on the Ettingshausen effect has a small performance energy ratio ε and small efficiency η_Σ with respect to the Carnot cycle. Calculations show that for a five-stage system $\varepsilon = 0.0019$ and $\eta_\Sigma = 0.1\%$ at $T_0 = 297$ K and $T_2 = 198$ K .

8.3. Adiabatic demagnetization method.

In 1926, Debye and Joucomb developed this method independently. Fig. 8.3 shows a plot of a **T-s** diagram for paramagnetic salt. For effective magnetic cooling the "non-magnetic" entropy of a system (lattice part s_l) should be small compared with magnetic entropy s_{0H} due to uncompensated electron spins. Usually s_l near 1 K is small. At these temperatures, the ion interaction energy is small compared with the thermal energy **kT**. Therefore, the spatial orientation of ions remains randow, and a "magnetic" spin entropy part is still essential.

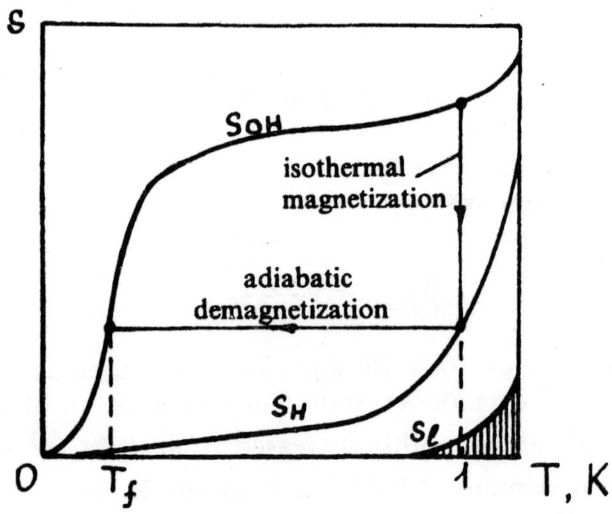

Fig. 8.3. T-s diagram for paramagnetic salt.

In Fig. 8.3 the line s_{0H} shows salt entropy as a function of temperature. With application of a magnetic field (about 1 T), a greater number of ions are aligned parallel to the field and entropy is markedly lower than in the absence of magnetic field (line s_H).

Thus, if magnetized salt is cooled to 1 K and the field is then sharply switched off, then under adiabatic conditions the salt temperature is decreased from 1 K to T_f (transition from s_H to s_{0H} at $s = const$).

At a temperature below that determined (θ is the characteristic temperature or Curie point), when interaction energy of election spins is comparable with the thermal **kT** or greater, the system randomness is eliminated owing to the interaction force affect, entropy strongly decreases and no longer depends on magnetic field. Therefore, salts at temperatures below θ are not yet effective as working substance of the cryogenic cycle.

8.3.1. Adiabatic demagnetization thermodynamics

The first law of thermodynamics when applied to substances in a magnetic field may be written as:

$$T\,ds = du + M\,dH\,, \tag{8.12}$$

where **M** is the magnetic moment; **H** is the magnetic field intensity.

In the majority of magnetic processes, all terms incorporating non-magnetic magnitudes (**p**, **v**, etc.) may be neglected. Then, based on the first and second laws of thermodynamics, for variable **T** and **H**, we have

$$ds = \frac{C_H}{T}\,dT + \left(\frac{\partial M}{\partial T}\right)_H dH\,, \tag{8.13}$$

where C_H is the substance heat capacity in a constant magnetic field **H**.

For constant entropy, from (8.13), we obtain

$$\alpha_{s,H} = \left(\frac{\partial T}{\partial H}\right)_s = \frac{-T}{C_H}\left(\frac{\partial M}{\partial T}\right)_H . \qquad (8.14)$$

From relation (8.14) it follows that under isoentropic demagnetization the cooling effect occurs as $(\partial M/\partial T)_H$ is negative for ordinary paramagnets.

The coefficient

$$\alpha_{s,H} = \left(\frac{\partial T}{\partial H}\right)_s$$

may be called the differential adiabatic demagnetization effect similar to differential effects under gas throttling (α_h) and isoentropic expantion (α_s).

To satisfy the Curie law

$$M = A H / T$$

(where A is the constant), a relation may be found for a finite demagnetization temperature:

$$T_f = T_0 \sqrt{1 - \frac{A H_0^2}{C_H T_0^2}} , \qquad (8.15)$$

i.e. cooling increases with magnetization field H_0.

8.3.2. Adiabatic demagnetization cooling device

Fig. 8.4 shows a schematic of adiabatic demagnetization cooling device. Paramagnetic salt is placed into vessel A submerged in liquid helium (1 K). First, the vessel A

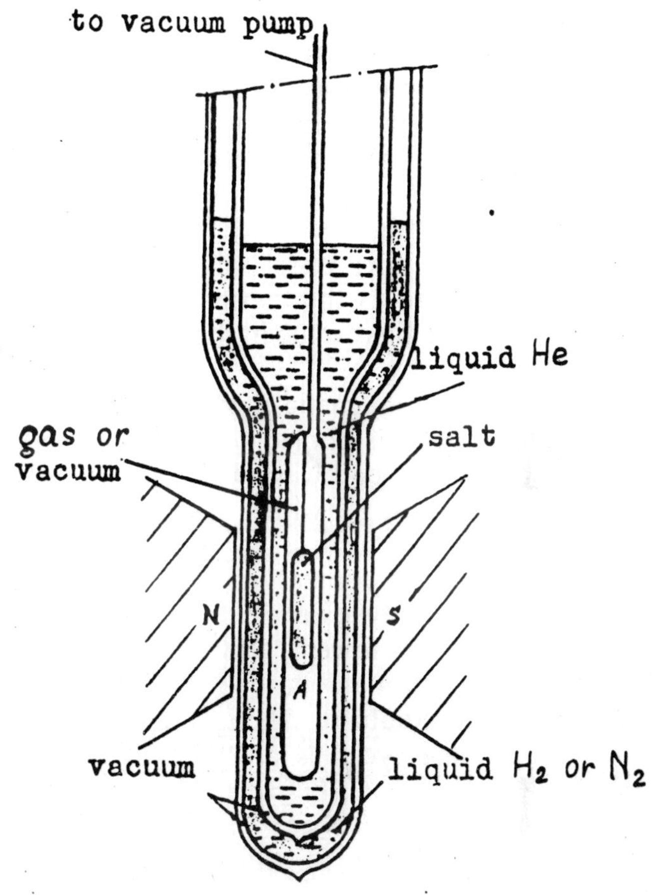

Fig. 8.4. Adibatic demagnetization cooling device.

is filled with gaseous helium. As a result, salt may be cooled to 1 K after application of a magnetic field. Then helium pumped from vessel A and salt appears under vacu-um insulation. With magnetic field switched off, salt is cooled. If salt is not tested, then the sample is placed into a salt-filled ampule or a good heat conductor is placed between it and the salt.

Cooling may be effected in two stages. In this case, two salt-filled vessels are located one above the other and connected via thermal pathway (by a superconductor). This thermal pathway is characterized by high thermal conductivity in a normal state, and low thermal conductivity in superconductive state.

At first stage magnetic field in the upper vessel is switched off and the upper vessel salt is cooled. The lower vessel salt is also cooled as the thermal pathway remains in normal state (magnetic field value in the lower vessel is higher than critical).

At second stage magnetic field is switched off in the lower vessel and its salt is cooled to lower temperature. In this case, the thermal pathway with no magnetic field becomes superconducting with specific low thermal conductivity. Therefore, heat is not transferred from the upper to lower vessel. In contrast to one-stage cooling, this method may use essentially lower field intensities.

At present, the lowest temperature obtained by adiabatic demagnetization amounts to 0.0014 K.

Based on the adiabatic demagnetization principle, a cycle machine (2-min cycle) is designed using periodically switched off thermal contacts between salt, sample and liquid helium bath. Termal contacts are made of superconductors, whose superconduction may, at will be destroyed by application of a magnetic field. Such a machine at 0.26 K has a refrigerating capacity of 0.84 mJ per cycle.

Recently, cooling units have been designed, based on the electrocaloric effect similar to the variety described above magnetocaloric. These units use adiabatic dielectric depolarization in an electric field instead of adiabatic demagnetization of magnets.

8.4. Adiabatic magnetization of a superconductor

Adiabatic magnetization of a superconductor may be employed to obtain ultra-low temperatures. At a temperature lower than critical T_{cr} for a given superconductor the entropy of substance in a normal state is the higher than the entropy in a superconducting state. Hence, with application of a magnetic field, the superconductor temperature will decrease.

8.5. Nuclear demagnetization

A cooling limit with adiabatic demagnetization of paramagnetic salts is caused by ion interaction. Gorter, and independently Curtie and Simon, assumed that adiabatic demagnetization of substances whose nuclei possess a magnetic moment may yield substantially lower temperatures. Particles constituting a nucleus also have spins, and if these are not compensated, then a nucleus will be characterized by a magnetic moment (e.g. copper, cobalt, etc.). These magnetic moments are approximately 1000 times smaller than those of paramagnetic salts, and therefore the slightest extraneous effect disturbs their orientation in a magnetic field.

The orientation may remain only at the lowest temperatures obtained preliminarily by adiabatic demagnetization of a paramagnetic salt (up to 0.01 K) or in dissolving refrigerants. This is an essential stage for nuclear cooling conditions.

Theory and experiment show that demagnetizing nuclei yield spin temperatures of about $10^{-5} \div 10^{-6}$ K. A system of nuclear spins alone is cooled to these temperatures but not the entire sample. Then, temperatures of a spin system and of a crystal lattice flatten ont.

However, this process is difficult and long. The entire sample may obviously be cooled to 10^{-4} K using nuclear demagnetization.

At present, devices based on demagnetizing paramagnetic salts $Gd_2(SO_4)_3 \times 8\ H_20$ are designed to attain higher temperatures (2 K and 10 ÷ 20 K) under cycling from tens of fractions Hz to several Hz.

A reciprocating piston-type device provides heat transfer from a level of 2 K to a level of 4 K at a refrigerating capacity of 0.52 mW to 1.2 W. Both a super-conducting magnet (providing a 5 T field) and sample with salt move in this device.

Another rotation-type device with a rotation frequency of working sample up to 10 Hz provides for heat transfer from a level of 2 K to a level of 10 K with a greater refrigerating capacity. The salt is magnetized to 5 T. A regime is possible with less than 1 s cycling for producing cold at a level of 10 ÷ 20 K. Also, electrocaloric devices are being designed for these temperature levels.

Questions

1. What does the Peltier effect mean?
2. Derive the formula for the coefficient of performance of a refrigerating machine of a thermobattery and optimum current.
3. How is the criterion for the efficiency of thermoelectric cooling calculated?
4. What does the Ettingshausen effect mean?
5. How does cooling according to the adiabatic demagnetization method occur?
6. Derive the formula for the differential effect of adiabatic demagnetization.
7. What are the specific features of the cooling process on nuclear demagnetization?

9

COOLING METHODS BASED ON SPECIFIC PROPERTIES OF HELIUM ISOTOPES

Helium is widely used in cryogenic systems: this applies to basic working substances at temperatures below 80 K. Helium (its most widely used isotope He^4) is usually employed in systems involving throttling, isoentropic expansion and other processes considered in Lectures 5÷7. However, these classical processes become inapplicable for cooling to temperatures below 1 K. At the same time, it appears possible to implement some other cooling effects, using specific properties of helium at ultralow temperature.

Both stable isotopes of helium (He^4 and He^3) form unique quantum fluids, i.e. fluids in which quantum effects manifest themselves on a macroscopic scale. This fact is the main reason why helium isotopes can be used to attain ultralow temperatures.

9.1. Obtaining low temperatures with dilution of He^3 in He^4

In 1951, H. London proposed a method of obtaining ultralow temperatures, based on use of the thermal effect when He^3 is diluted in He^4. This process is similar to ordinary liquid evaporation; in this situation, pure condensed He^3 serves as a liquid phase while a weak He^3 solution in He^4, serves as a vapor. Heat absorption during dilution becomes possible thanks to the specific nature of interacting particles of these liquefied isotopes.

Throughout the considered temperature range (below 1 K), liquid He^4 is in a superfluid state and possesses almost zero viscosity and entropy. It is obvious that this fluid is inert in its hydrodynamic and thermal aspects. Such behavior is not typical of He^3. In diluting He^3 in He^4, the atoms of these fluids do not interact because of the inertness of He^4; hence, the He^3 transition to a He^4-filled volume where the concentration of He^3 atoms is small is similar to its expansion and is accompanied by energy absorption. In this case, in the solution a gaseous phase of He^3 atoms appears for which He^4 is only a "supporting" medium.

Thus, dilution changes the orderliness of a system; as a result of fluid-gas transition, the He^3 entropy increases and heat capacity also increases. To a first approximation, for a weak solution of He^3 in He^4,

$$\mathbf{c_v = \frac{3}{2}R}$$

(ideal monatomic gas); at $\mathbf{T} < 0.4$ K, this value of heat capacity starts decreasing. Dilution is accompanied by heat absorption on transition from one phase (pure He^3) to another (weak He^3 solution) as during ordinary evaporation.

The properties of He^3 and its weak solution have been well studied; the fullest theory takes account of He^3 particle interaction and is based on the model for a weakly interacting gas, whose particles obey Fermi-Dirac statistics.

The natural (spontaneous) process of phase separation is another important property of He^3 - He^4 solution, i.e. there spontaneous separation occurs at temperatures below 0.827 K. Atoms of the heavier superfluid He^4 are released from the solution, accumulated in the bottom of the vessel.

A clear interface is formed between liquid He^3 and He^4. In the **T-x** liquid state diagram (Fig. 9.1a), the separation region is limited by the so-called separation curve. On cooling a solution with initial concentration **x**, on line 1-2 (Fig. 9.1a) it separates and disintegrates into two phases: normal **(n)** and superfluid **(s)**. The He^3 concentration **x** sharply increases in the normal phase $\mathbf{x_n}$ and decreases in the superfluid phase $\mathbf{x_s}$. Values of concentrations are found from points 3 and 4, where the isotherm passing through point 2 intersects the separation curve. The normal-to-superfluid phase ratio is determined by the section ratio:

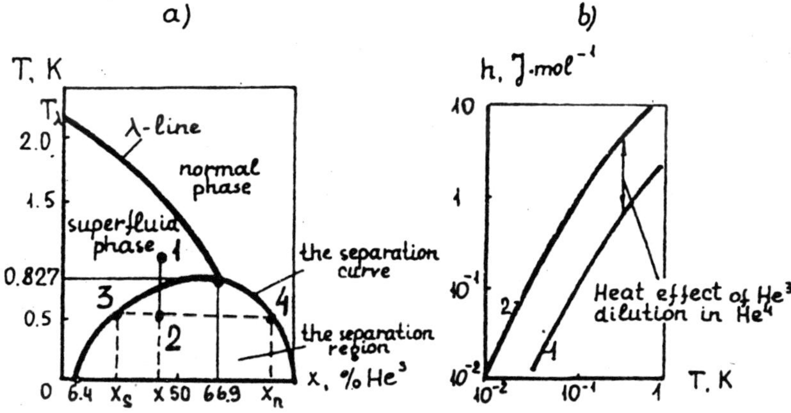

Fig. 9.1. Diagrams illustrating the properties of He^3 diluted in He^4:
a, T-s diagram of liquid states of a system of He^3 and He^4;
b, pure He^3 (curve 1) enthalpy and partial He^3 enthalpy in solution (curve 2) as a function of temperature.

$$\frac{m_n}{m_s} = \frac{x - x_s}{x_n - x} \cdot$$

At $T \rightarrow 0$, the abscissa value of the separation curve is equal to 6.37%, a minimum dilution of He^3 in He^4. Transition of He atoms from the upper (n) to lower (s) phase through their interface also appears like liquid He^3 evaporation.

The first successful experiments on He^3 dilution in He^4 were carried out in 1964, and the first effective device was designed in 1965.

The principles of a dilution refrigerator are shown in Fig. 9.2. Almost pure gaseous He^3 from the vacuum pump at $p = 66.5 \div 80$ kPa is directed to the forward flow. Gas is successively cooled in a bath filled with liquid nitrogen and helium (not shown in the diagram). It then condenses in a bath filled with He^4 boiling at $T \sim 1$ K. Further, He^3 in liquid

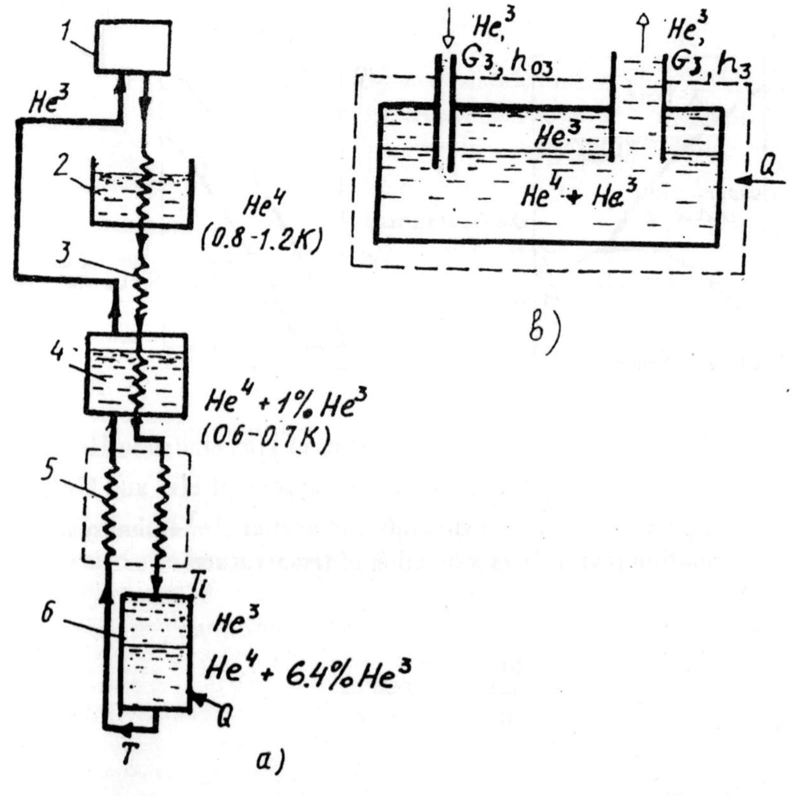

Fig. 9.2. Schemes of a refrigerator (a) and mixing bath (b) operating when He^3 is diluted in He^4: 1, vacuum pump; 2, He^4 bath; 3, capillary; 4, evaporation bath; 5, heat exchanger; 6, mixing bath.

phase passes through a capillary where its pressure degrades (a capillary is responsible for He^3 flow-rate). After this, the He^3 flows into the evaporator coil ($T_v \approx 0.6 \div 0.7$ K), is cooled, goes to a heat exchanger where it is cooled by reverse flow to a remperature T_i and enters mixing bath 6.

In the mixing bath, phase separation takes place; almost pure He^3 is found in the top part while a weak He^3 - He^4 solution concentrates in the bottom part. A He^3

absorption; under adiabatic conditions this decreases the temperature while under isothermal conditions it provides a refrigerating capacity **Q**. From the mixing bath reverse He^3 flow is directed through a heat exchanger to the evaporation bath.

In this case, He^3 in the heat exchanger moves through almost a stationary column of He^4, thus cooling the forward flow entering the mixing chamber.

The evaporator temperature is higher than the mixer value, and the corresponding equilibrium He^3 concentration constitutes 1%; however, a partial He^3 pressure at an evaporator temperature of 0.6 K is approximately 1000 times higher than that of He^4; therefore, almost pure He^3 evaporates from the bath. A He^3 evaporator vapor pressure of about 0.60 Pa is provided by the vacuum pump. Thus, a continous closed refrigerator cycle takes place. A temperature level in the mixing chamber is usually kept at 0.1 to 0.01 K at relatively high heat load **Q**.

To determine the refrigerating capacity **Q**, the enthalpy ($J \cdot mol^{-1}$) difference of the He^3 flow at the mixing bath inlet and after dilution must be known. Then, the amount of heat absorbed during dilution, W, is

$$\mathbf{Q} = \mathbf{G_3} (\mathbf{h_3} - \mathbf{h_{03}}) , \qquad (9.1)$$

where $\mathbf{G_3}$ is the flow-rate of circulating He^3, $mol \cdot s^{-1}$.

For the dilution process the enthalpy may be found from the expression $\mathbf{dh = T\ ds}$. The entropy of the Fermi gas, He^3, is a linear function of temperature: $\mathbf{s = a\ T}$. From this fact it follows that the enthalpy of both pure He^3 and its solution is a quadratic function of temperature:

$$\mathbf{h_{03} = 12\ T_i^2\ ; \qquad h_3 = 94\ T^2 .}$$

These relations for both He^3 phases are plotted in Fig. 9.1b. In accordance with expression (9.1), the dilution refrigerating capacity is

$$Q = G_3 (94\,T^2 - 12T_i^2)\,. \qquad (9.2)$$

For high efficiency of heat exchanger 5 (Fig. 9.2), the temperature difference $T_i - T$ is very small and $T_i / T \approx 1$; in this case, heat load ($J \cdot mol^{-1}$) is maximum and equal to

$$\frac{Q}{G_3} = 82\,T\,. \qquad (9.3)$$

Also, formula (9.2) implies that $Q = 0$ at $T / T_i = 0.36$, i.e. in this limiting case the inlet flow may decrease its temperature as many as approximately three times. Thus, possible temperature range is $T / T_i = 0.36 \div 1$. The difference between T_i and T depends on the effeciency of heat exchanger 5; obviously, the greater the difference between T_i and T, the smaller efficient heat load Q.

In a limiting case, at $T/T_i = 0.36$ and mixing bath temperature $T = 0.01$ K :

$$(T_i)_{max} = \frac{0.01}{0.36} = 0.028\ K\,,$$

i.e. $Q \to 0$ for a temperature difference on the cold heat exchanger end

$$\Delta T_{max} = 0.028 - 0.01 = 0.018\ K\,.$$

From this, it follows that heat exchanger 5 must possess a high efficiency, thus providing a very small flow temperature difference. At the same time, heat transfer at such low temperatures proceeds at a low rate.

The Kapitza thermal resistance $R_k = (T_w - T_l) / Q$ exerts a determining influence on heat transfer. Axial heat conduction in a heat exchanger, ambient heat flux, and viscous heating in He^3 flow, are other losses affecting process efficiency. The loss effect results in a real cooling effect at $T = 0.1$ 3-5 times smaller than the theoretical value. In the existing dilution refrigerators, a He^3 flow circulation rate is $G_3 \leq 10^{-4}$ mol·s⁻¹. The refrigerating capacity of such refrigerators usually amounts to 0.2 mW at $T = 0.1$ K and decreases to 0.01 mW at $T = 0.04$ K. A minimum attained temperature is 0.003 K in single-pass systems and 0.05 K in a continuous process.

There is no doubt that dilution refrigerators will be constantly improved and have wide used in research. The continuous action of these systems, relatively high refrigerating capacity and simplicity in providing thermal contact between a fluid and an object to be cooled represent their advantages over adiabatic demagnetization. There are reasons to assume that this method enables one to use ultralow temperatures for engineer purposes.

9.2. Obtaining low temperatures at adiabatic He^3 crystallization

In 1950, I. Pomeranchuk proposed a new cooling method based on specific properties of the He^3 isotope (Pomeranchuk effect). The melting curve $\mathbf{p = f(T)}$ showing the equilibrium states of He^3 liquid and solid states is quite different in nature. At a temperature of 0.32 K and an equilibrium pressure of 2.9 MPa, the curve for He^3 melting has a minimum (Fig. 9.3). It is evident that elevating the pressure in this two-phase system at $T < 0.32$ K reduces the temperature (process AB) whereas the normal behavior of bodies is associated with their heating under compression. He^3 behavior of this type has a quantum nature; it results from the strong influence of zero energy and is attributed to the character of orientations of magnetic moments of nuclei.

The specific properties of He^3 also depend on the value of its entropy in liquid and solid phases. In ordinary systems, the fluid entropy always exceeds that for the crystal

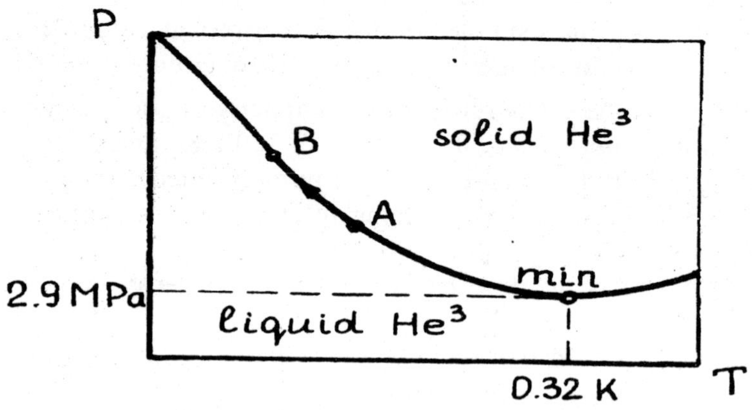

Fig. 9.3. Curve for He^3 melting at low temperature.

since in a crystal there is a higher degree of orderliness. In He^3, at **T** < 0.32 the solid phase entropy is greater than that of the fluid which possesses greater order. As a result, the fluid-crystal transition due to compression at **T = const** is accompanied by increasing the entropy and absorption of heat **Q**. On the other hand, this transition under adiabatic conditions **(s = const)** is accompanied by a decrease in temperature.

Fig. 9.4 shows a **T-s** diagram for He^3 at **T** < 0.5 K and cooling due to adiabatic compression. Cooling proceeds in two stages:

(a) preliminary liquefaction of He^3 and its cooling to a temperature below 0.32 K, providing a state with lower entropy than in the solid phase at the same temperature (point A);

(b) He^3 crystallization by isoentropic fluid compression: process A-B; as a result, the temperature is decreased by $\Delta T = T_i - T_f$. The amount of heat absorbed under isothermal compression (process A-C) may be determined as a phase transition heat $Q = T\,(s_s - s_l)$.

The solid He^3 entropy at T < 0.32 K is approximately constant:

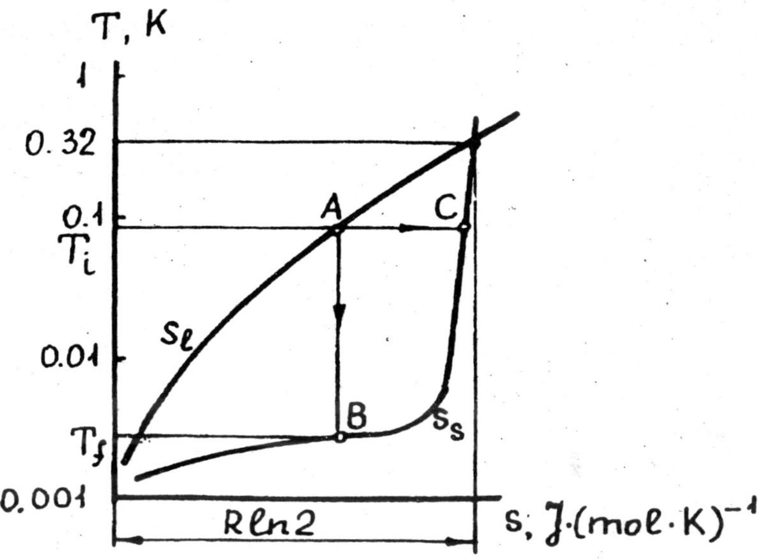

Fig. 9.4. He^3 entropy diagram on the melting curve for liquid (s_l) and solid (s_s) phases.

$$s_s \approx R \ln 2.$$

The behavior of the fluid entropy is close to linear

$$s_l \approx 18\,T.$$

Then the amount of the absorbed heat, J·mol^{-1} is

$$Q = T\,(R \ln 2 - 18\,T)\,. \qquad (9.4)$$

The temperature variation ΔT may be seen directly by the T-s diagram. The final temperature T_f to be attained under compression is determined by the initial T_i: the lower T_i, the smaller is T_f (Fig. 9.4).

The solid phase fraction in relation to the total helium

mass in a chamber $z_s = m_s/m$ may be easily calculated at any point of process A-B, assuming that both helium phases (solid and liquid) are all the time at thermal equilibrium, and the heat capacity of a chamber itself is negligibly small.

For the process with $s = \mathrm{const}$, the clear relation is valid

$$m\, s_l(T_i) = m_s\, s_s(T) + (m - m_s)\, s_l(T)\,,$$

from which we obtain

$$z_s(T) = \frac{s_l(T_i) - s_l(T)}{s_s(T) - s_l(T)}\,.$$

It is easy to see that at a given temperature T the solid phase fraction z_s in a chamber depends on initial temperature T_i so that with decreasing T_i, $z_s(T)$ decreases too.

In principle, this method of cooling is single-pass. However, the compression process ceases at $z < 1$, and the chamber temperature will remain constant unless external heat fluxes, bring about crystallization of the entire remaining fluid. For good thermal insulation, the temperature can remain constant for several days.

The main practical difficulty in implementing Pomeranchuk's method is the avoidance of heating due to friction during compression of the He^3 chamber. It may be shown that it is sufficient for just 1% of the mechanical work done with a system under compression to be converted into heat due to friction for cooling to be zero. This fact explains why for a long time after I. Pomeranchuk proposed the new cooling method, no experimentalist even tried to check it.

Only in 1965 Yu. Anufriev succeed in cooling by Pomeranchuk's method. Later on, other successful experi-ments were caried out, in which temperatures of about 1 mK were attained.

Pomeranchuk's method is in use only for cooling He^3 itself and for studying its properties. Since in this case the investigated object also serves as a cryogenic agent, the complex problem of providing thermal contact between them is eliminated. In addition, certain dufficulties may arise from the presence of a heterogeneous solidfluid mixture in the cnamber which inevitably appears in the cooling process.

Use of He^3 for cooling other objects is restricted by the fact that a solid phase of He^3 is formed in the hottest parts of the chamber (He^3 crystallization needs heat). Therefore, a cooled object which is always slightly hotter than its surroundings starts to take on a covering of solid He^3, thereby sharply deteriorating the thermal contact and heat transfer conditions. However, it is very probable that this problem will be solved very soon by using an applied magnetic field. As we have seen, solid He^3 is formed in those parts of the chamber where magnetic field induction is largest.

Therefore, a high possibility arises of cooling an object by concentrating an applied field in other parts of the chamber. This development of Pomeranchuk's method is very promising for various physical experiments over temperature range 2 to 10 mK, i.e. just in that region of temperatures above which the dilution of He^3 in He^4 is used successfully and below which there is nuclear demagnetization.

9.3. Cooling by Kapitza's mechanocaloric effect

In discussing properties of liquid He-II, it should be noted that its behavior is qualitatively well explained within the framework of the "two-fluid" model providing for the existence of "normal" and "superfluid" components in fluid. Since the superfluid component viscosity is equal to zero, it moves freely in He-II through a narrow capillary. At the same time, the normal component cannot move through this capillary and is filtered out. As a result of this process, He-II separates into two fluids: one is rich in superfluid component and another mainly contains the normal component. This effect called mechanocaloric is accompanied by a decreasse in temperature through the fluid capillary.

Fig. 9.5 illustrates the mechanocaloric effect. As the superfluid component entropy is equal to zero, a thin capillary **C** must act as an "entropy filter" not transmitting the viscous normal component.

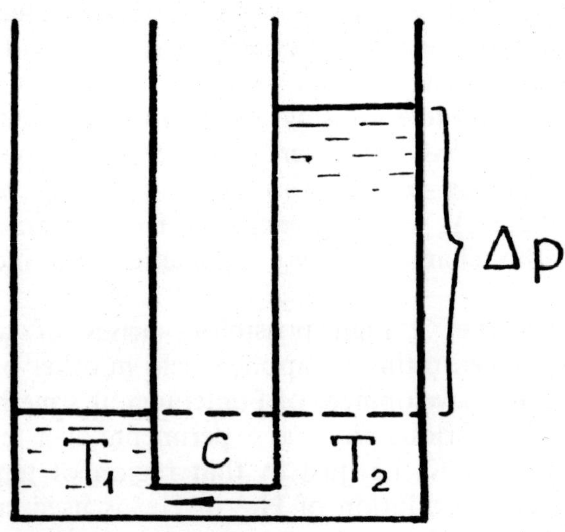

Fig. 9.5. Scheme illlustrating the mechanocaloric effect in liquid He-II.

The observed temperature decrease after the filter is determined by the fact that the entropy of superfluid that has transmitted through the filter is equal to zero, and hence its temperature is lower. P. Kapitza was the first to discover this effect and obtain experimental support. Difficulties associ-ated with using this effect for cooling are determined by the small heat capacity of the superfluid component of He-II and by the fact that this fluid cannot serve as a heat carrier.

Questions

1. What does the thermal effect of He^3 diluted in He^4 mean?

2. What are the specific properties of the He^3 - He^4 solution?

3. How is the lamination of the He^3 - He^4 solution calculated using the **T-s** diagram?

4. Derive the formula for the refrigerating capacity for He^3 diluted in He^4.

5. What are the distinctive features of the He^3 melting curve?

6. What are the specific features of the **T-s** diagram for He^3 at $T < 0.32$ K?

7. What is the mechanism of cooling according to the Pomeranchuk method?

8. How is the solid phase fraction of adiabatic He^3 crystallization calculated?

9. What are the difficulties in using the Pomeranchuk method in practice?

10. What does the cooling method using Kapitsa's mechanocaloric effect mean?

10

HEAT TRANSFER AT LOW TEMPERATURES

Processes of heat removal from bodies to be cooled, and of heat given out to the environment or intermediate agents, are associated with the heat transfer phenomenon.

In the walls of the elements of a cryogenic systems heat is transferred by heat conduction. Most useful are unsteady heat conduction processes that take place on cooling or heating of a system as well as steady ones, due to which heat flows from the ambient medium via thermal insulation, supports and pipelines.

A distinctive feature of heat-conduction process at low temperatures (especially, in the cryogenic region) is that thermophysical properties of structural materials depend strongly on temperature.

On the wall surface of a system, heat exchange processes occur (heat exchange between solid wall and liquid, or gas, or two-phase medium). Heat exchangers perform heat transfer processes (heat transfer from one medium to another via a solid wall) in which the mechanisms of heat conduction and heat exchange are combined. A specific feature of a heat exchange process is the motion of a medium where heat propagates. Heat exchange in this case occurs by transfer of macroscopic masses of a liquid or a gas (convective heat exchange).

In refrigerating and cryogenic systems, there is both free convection (gravitational convection), where the non-uniform density distribution is responsible for the medium motion, and forced convection, where the medium moves owing to external impacts (for example, pump or excess pressure in the flow-rate tank).

Convection may occur both in single-phase (liquid, gas, medium in near-critical state) and two-phase (vapor-liquid) media. In the latter, heat exchange is accompanied by a phase transition (boiling or condensation). Depending on the heating surface temperature and some other factors, one of three main boiling regimes (nucleate, transition and film) and one of two condensation regimes (film or droplet) may occur.

In two-phase (vapor-liquid) flow at low temperatures, flow regimes, other than high-temperature channel flows are predominant. For example, in cryogenic piping vapor moves more frequently near a wall while a liquid moves in the channel center. In steam generators, liquid moves at the wall and vapor at the channel center.

10.1. Boiling of cryogenic liquids

10.1.1. Nucleate boiling

Nucleate boiling occurs writh relatively low temperature heads (temperature head $\Delta T_w = T_w - T_s$ where T_w is the wall temperature, and T_s is the saturation temperature corresponding to a cavity or channel pressure). In nucleate boiling, vapor is formed as individual bubbles which periodically appear on the superheated wall, rise and after departure from the wall move in the liquid. High heat exchange intensity is provided both by turbulization of the thermal boundary layer of the liquid on nucleation on the wall, and also by liquid mixing due to bubble rise.

Density fluctuations due to the thermal motion of molecules initiate nucleation in the superheated liquid. To achieve a stable existence of the formed bubble the vapor pressure force inside it must be no less than the sum of the pressure forces of liquid and surface tension (otherwise the bubble collapses). Hence, it follows that for a stable vapor nucleus to be formed, a liquid layer adjacent to a wall must be superheated to $T_{sup} > T_s$, where the saturated vapor pressure in a randomly formed bubble is sufficient to prevent bubble collapse.

The recesses of the roughness of the heating surface

(scratches, cracks, etc.) and the deteriorated wetting regions are the most probable vapor nucleation sites.

Bubble growth occurs because of the heat of the superheated liquid and the heating surface supplied to the bubbles. This heat is used in liquid evaporation and expansion work.

With increase of vapor bubble radius, the lift force increases, with a tendency for the bubble to leave the wall. The bubble departure diameter is determined as the largest one for which the equilibrium of the forces acting upon a bubble is still satisfied. At $\mathbf{D > D_{dep}}$, the forces that keep a bubble near the wall (surface tension and inertia reactoin of the liquid displaced from the wall) may not balance the action of the forces that move a bubble off from its nucleation site (buoyancy force; in the forced flow dynamic heat of liquid flow). In nucleate boiling, the heat flux density is composed of two components, one of which is connected with liquid evaporation near the heating surface, the other with convective heat exchange due to liquid mixing by rising and departing bubbles near the heating surface.

10.1.2. Film boiling

Film boiling arises at fairly high values of temperature head when the wall is separated from the liquid by a vapor layer. The interface may be smooth and stable or wave and oscillating, depending on the relationship between the temperature heat and subcooling $\mathbf{\Delta T_{sub} = T_s - T_l}$ where $\mathbf{T_l}$ is the bulk liquid temperature.

Heat flux removed from the wall is transferred through a vapor film to the interface where it is used in liquid evaporation and heating. The process is self-regulated so as to ensure a vapor-film temperature drop to the saturation temperature $\mathbf{T_s}$ for the interface at any wall temperature $\mathbf{T_w}$. Heat transfer is specified by the conditions of vapor removal in the mass force field and hydrodynamics of interface oscillations with regard to the vapor formation on it.

Interface oscillations are caused by the hydrodynamic instability of a two-phase system, in which the vapor is at the side or at the bottom of the liquid.

For constant subcooling, the rate of interface oscillations increases with temperature head. This is attributed to increasing the temperature gradient near the heating surface that produces more intensive evaporation on the wave crests. The increase of the oscillation amplitude yields velocity components normal to the wall and turbulent transfer enhancement in the vapor and liquid. Thus, on subcooled liquid boiling the total heat flux as the heat flux used on liquid heating increases with temperature head.

At constant temperature head, the increase in subcooling decreases the vapor film thickness and the rate of the interface oscillations. This is explained by the fact that the heat flux spent on liquid heating through the vapor film increases with subcooling. Hence, the temperature gradient in the vapor film increases and (for $\mathbf{T_w = const}$) its thickness decreases. Damping of interface oscillations is caused by the fact that with increased subcooling more intensive vapor condensation in wave troughs occurs. Thus, for fairly high subcooling the interface becomes practically smooth.

For film boiling in channels, heat exchange and hydraulic resistance depend substantially on the film boiling regime.

There are following regimes of film boiling in channels:

(1) Rod flow regime - the liquid saturated or subcooled moves in the form of turbulent or laminar jet ("liquid" rod) in the central part of the channel.

(2) Plug flow regime - this regime originates from the rod variety by the action of developing capillary waves on the liquid surface or by the action of the inertia discontinuity of the liquid "rod" into individual "plugs" with pulsation in flow rate.

(3) Dispersed flow regime - the liquid moves in the form of drops distributed in the superheated vapor flow. This

regime originates from the first two owing to the dynamic effect of the vapor on the liquid "rod" or liquid "plugs".

If a channel is located at an angle to the direction of the mass force field, then the phase distribution in the above flow regimes will be asymmetric, especially for small Reynolds number of the mixture. This may result in the onset of a new "stratified" film boiling regime. It originates from the rod type, for example in a horizontal tube.

Liquid separated from the walls with the vapor film which moves upwards flows in the lower part of the tube. The vapor concentrated in the upper part of the tube moves along the flow axis. If the vapor velocity with respect to the liquid is quite high, then the waves on the liquid surface may approach the upper wall of the tube.

10.1.3. Transition boiling

In the intermediate region $\mathbf{T_w}$, where nucleate and film boiling are unstable, a transition boiling regime arises when nucleate and film boiling are in periodic exchange for each section of the heating surface.

The heat flux density as a function of temperature heat is shown in Fig. 10.1.

According to the existing classification of boiling regimes, the transition boiling region is intermediate between nucleate and film boiling regions, and its boundaries are considered to be the extrema of the boiling curve $\mathbf{q(\Delta T_w)}$, viz. the points of the nucleate boiling crisis ($\mathbf{\Delta T_{cr1}}$, $\mathbf{q_{cr1}}$) and film boiling crisis ($\mathbf{\Delta T_{cr2}}$, $\mathbf{q_{cr2}}$). The main specific feature of the boiling curve at temperature heads ($\mathbf{\Delta T_{cr1} \div \Delta T_{cr2}}$) is its nega-tive slope:

$$\frac{\mathbf{dq}}{\mathbf{d(\Delta T_w)}} < \mathbf{0} \,.$$

Under steady boiling for the prescribed constant heat load ($\mathbf{q = const}$), this fragment of the boiling curve is not

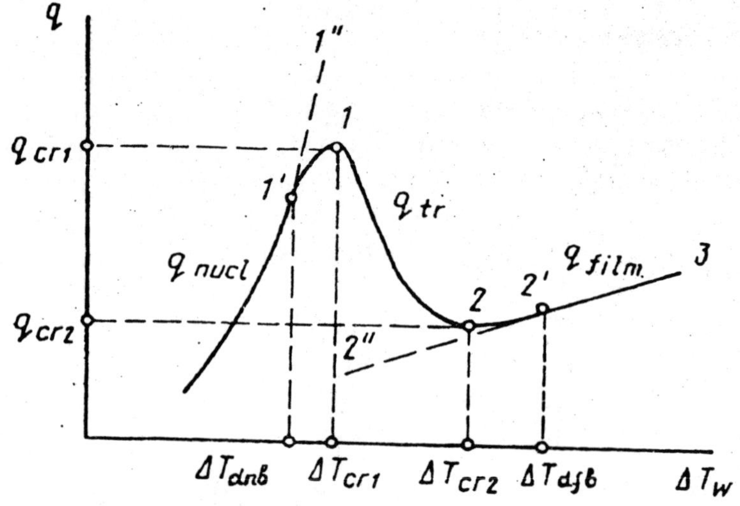

Fig. 10.1. Transition boiling region boundaries:
1, 2, boundaries of the fragment of the boiling curve with a negative slope; 1', boundary of the nucleate boiling tability; 2', boundary of the film boiling stability.

realized. Only at boiling crisis when wall temperature sharply increases (1st crisis) or decreases (2nd crisis) does the unsteady process cover the range of temperature heads (ΔT_{cr1}, ΔT_{cr2}). In this case, the heat exchange intensity in this transition boiling region specifies the rate of wall-temperature variation.

Another approach is also possible for determining the transition boiling region. This is based on the fact that in this region neither nucleate nor film boiling possesses stability (for either hydrodynamic or thermodynamic reasons). In this case, points 1' and 2' (Fig. 10.1), at which nucleate and film boiling regimes start degrading stability, should be considered as transition boiling boundaries. Point 1' ("dnb" - deviation from nucleate boiling) is characterized by the formation of unstable local dry spots on the heating surface with rise of temperature head T_w (or by their disapearance on decrease of T_w). Point 2' ("dfb" - deviation from film boiling) is characterized by the formation of unstable local cold spots on the heating surface with decreasing T_w (or their disappearance with increasing T_w).

The shortcoming of the choice of points 1' and 2' as the boundaries of the transition boiling region is associated with the asymptotic behavior of the transition boiling curve near these boundaries which results in practical difficulties in exact determination of the position of points 1' and 2' using experimental data (whereas the extremum points 1 and 2 are easily fixed).

Analysis of results from observation, high-speed photography and filming of transition boiling shows that in the transition boiling region wetted and dry sections on the heating surface co-exist at each time instant, and on an extensive heating surface each wall section appears to be alternately wetted or dry.

Dry spots on the heating surface originate in the nucleate boiling region at some temperature head $\Delta T_{dnb} < \Delta T_{cr1}$. In the presence of liquid subcooling, the difference $(\Delta T_{cr1} - \Delta T_{dnb})$ may increase substantially.

Observations show that the vapor film in the transition boiling region is hydrodynamically unstable. As a result, the liquid shoots out from time to time from the heating surface.

Other liquid behavior is specified by the temperature head: in the low-temperature zone of the transition boiling region in the vicinity of the 1st crisis, where the liquid contacts the wall, nucleate boiling initiates and its specific features manifest themselves in drying of amicrofilm under a rising bubble. For higher values of temperature head, the liquid wets the wall for a short period of time and bubbles up explosively. Finally, for still higher temperature heads the liquid is directed to the heating surface but, not having approached it, is directed backwards because of intensive evaporation. In this latter case, unstable film boiling occurs, characterized by heat transfer higher than for a stable film.

On further increase, the vapor film ceases to collapse and stable film boiling starts. When the temperature heat decreases, the above processes are repeated in the opposite sense.

It is found that when the liquid is insufficiently purified, more intensive contamination of the heating surface occurs in the transition boiling regime compared with the nucleate regime.

This supports the fact that the liquid contacts the wall in the transition boiling region and also that in transition boiling the liquid film on the heating surface completely dries out, leaving scale whereas in nucleate boiling the wall layer is impregnated with the liquid. As a result, the majority of the admixture remains in the liquid.

The nature of the transient boiling process depends strongly on liquid subcooling up to a saturation temperature. On boiling of the saturated and weakly subcooled liquid there is a relatively thick vapor film with local discontinuities. On increased subcooling, the film thickness decreases and the number of its discontinuities increases. In passing some threshold value, the boiling nature sharply changes. On the heating surface regions occupied by extremely small vapor sites ($0.2 \div 0.3$ mm) appear, whose general appearance is of a dull surface. In this regime referred to as "explosive" or "cavitation", transition boiling is accompanied by specific high-intensity noise. Threshold subcooling, where a transition to cavitation boiling appears that decreases with increase in temperature head and decrease in heating surface roughness.

On transition boiling, some part of the heating surface is wetted by the liquid at each time instant and the remainder is covered with vapor film, each point of the heating surface alternately contacting either the liquid or the vapor phase of a boiling medium. The mean time of contact between the heating surface and liquid depends on the parameters of the process (temperature head, subcooling, etc.) as well as on the properties of the boiling substance and wall material. Since the intensity of heat transfer to the liquid is higher than that to the vapor, the processes at sites of contact between wall and liquid are predominant in transition boiling.

With regard to the mechanism of heat exchange on the wetted part of the heating surface, it is necessary to distinguish three zones in the transition region:

(1) low temperature heat zone near the region of the nucleate boiling crisis, in which the time of the contact between liquid and wall is fairly high, and nucleate boiling occurs at the contact site;

(2) high temperature head zone near the region of the film boiling crisis, where for small contact time nucleate boiling cannot develop, and where heat transfer from the wall to the liquid due to insteady heat conduction dominates;

(3) mean temperature head zone, in which contributions of nucleate boiling and unsteady heat conduction are comparable.

On increase of temperature head, the time of periodic contacts between liquid and wall decreases, and heat exchange intensity falls from high values typical of nucleate boiling to low values typical of film boiling.

The disturbance of the hydrodynamic stability of a vapor film and conservation of the thermodynamic stability of the liquid at the contact site are the necessary conditions for contact between liquid and wall. Consider how these conditions are satisfied when film boiling is replaced by transition boiling in the case of a slow decrease in the wall temperature.

It is obvious that the stable equilibrium of the vapor-liquid interface in a gravitional field is possible only when a less dense phase is located above a denser one (film boiling under the lower side of a horizontal heating surface). In all other cases (film boiling occurs on vertical, inclined, spherical surfaces), the interface is unstable since a heavier phase is above or on the side of a light phase.

The instability causes transverse motion of the interface (oscillating or aperiodic). However, for high temperature heads when the thickness of a vapor film is large, the liquid does not approach the wall. With decreasing temperature head, the vapor film becomes thinner, amounting to values comparable with the size of the oscillating amplitude of the interface, and the liquid is able to contact the wall, from the point of view of hydrodynamics. Whether this contact would occur in reality now depends on the thermodynamic conditions or on their combination with hydrodynamic conditions. When a wave crest approaches the wall in the superheated vapor region, the intensive evaporation on the wave crest and reactive forces developed may throw the liquid off from the wall, and contact will not take place.

If contact between liquid and wall having different initial temperatures still occirs, then some intermediate temperarature T_b arises at the contact boundary. If this temperature exceeds the limiting temperature of metastable liquid superheating T_{lim}, then explosive boiling-up of the thinnest layer takes place and the liquid is thrown off from the heating surface.

If $T_b < T_{lim}$, then the thermodynamic condition of wetting the wall with the liquid is satisfied, and fairly long contact arises between liquid and wall, accompanied by a strong increase in heat removal due to unsteady heat conduction of the liquid. With longer even contact time longer, this contact occurs because of nucleate boiling.

Thus, of the two necessary conditions for liquid-wall contact the one that is satisfied with a lower temperature head predominates. In particular, for boiling beneath a horizontal heating surface when the interface is more stable, it is expected that film boiling moves down in the region of low temperature heads, and the film boiling crisis may be hydrodynamic in nature. On boiling above a horizontal heating surface and on vertical, cylindrical and spherical surfaces, possible contact between the liquid and the wall is limited by the conditions of the thermodynamic stability of the liquid.

A transition boiling mechanism is associated with the onset of cycles and with the cessation of contacts between liquid and heated wall.

For high wall temperature T_w, the liquid is separated from the heating surface with a vapor film (stable film boiling region).

With decreasihg T_w, the vapor film becomes thinner, and the developed oscillations of the interface may force liquid to contact the heating surface. At the site, where the hot wall contacts the liquid, the liquid is heated, and stable vapor phase nuclei are formed after the liquid layer near the wall is superheated to a certain value. Then, the vapor bubbles rise, coalesce into a continuous film and separate the liquid from the heating surface. The vapor film formed in this case appears to be hydrodynamically instable, resulting then in contact between liquid and wall, and the process becomes cyclic. As the wall temperatuture is decreased, the contact time between liquid and heating surface increases, and the intensity of heat exchange in the transition boiling region increases from the low values typical of film boiling to high ones typical of nucleate boiling. On further decrease in the heating surface temperature, the number of vapor nuclei forming decreases to such an axtent that the rising vapor bubbles attain a departure size before they coalesce, and transition boiling is replaced by stable nucleate boiling.

10.2. Condensation at low temperatures

Vapor condensation on a cold wall may occur under film or drop boiling conditions, i.e. condensate is cooled either by a solid film or by separate drops. Film condensation appears on the surfaces wetted with condensate, while drop condensation appears on the non-wetted surfaces. Since cryogenic liquids possess very high mettability (wetting angle for the known construction materials close to zero), these are typical of film condensation.

On film condensation on the vertical wall in its upper part where the liquid film thickness δ and the mean velocity

w are small laminar flow at a smooth interface is observed. While the thickness of the film grows, waves develop on its surface. At Reynolds number

$$\mathbf{Re} = \frac{w\,\delta}{\nu_l} > 400 \ ,$$

the film flow becomes turbulent.

Since transition from vapor to liquid phase is accompanied by increase in density ($\rho_l > \rho_v$), vapor suction through the interface will occur on condensation, together with the flowing in of new vapor from the bulk.

Heat released on vapor condensation is transferred to the cooling surface through a liquid film. In this case, the total thermal resistance is composed of three components: thermal resistance of the liquid film, of the vapor and of the interface due to an interface temperature jump. For non-metallic liquids, including cryogenic ones, the last two components may be neglected (except for the case of high, vacuum).

For forced vapor flow along the surface, on when condensation takes place, frictional force acts on the liquid film on the vapor side. If the vapor moves downward, then the condensate velocity increases, the film thickness decreases and the heat transfer coefficient increases. If the vapor moves upward, then the liquid velocity decreases, the film thickness increases and the heat transfer coefficient decreases. If in the case of the ascending vapor motion the frictional force exceeds the force of gravity, then the condensate film will move upwards and the heat exchange coefficient will increase with vapor velocity. On condensation of the vapor moving inside the tube when the condensate flow is turbulent, the tube orientation does not affect the heat exchange intensity, i.e. the dynamic effect of the vapor on the condensate film predominates over the action of the gravitational force.

10.3. Heat exchange in near-critical region

In low-temperature systems, the refrigerant may be in a state close to the critical point, near which the properties of substance are strongly affected by temperature and pressure. The specific features of heat exchange processes in the substance in the state close to the critical point are attributed to a sharp change in thermodynamic and transport properties in passing to the critical temperature T_c. The specific heat c_p varies especially strongly and has a pronounced maximum at $p > p_c$ at some temperature T^* close to T_c. Moreover, experiments show that near the critical point the relaxation time (transition of the thermodynamic system from non-equilibrium to equilibrium state) increases. This region may therefore be characterized by non-equilibrium phenomena, for example by hysteresis of the relation $\rho(T)$ on heating under constant pressure and subsequent cooling (owing to non-equilibrium, the density ρ depends not only on pressure and temperature but also on process history).

A large temperature drop between the ambient medium and cryogenic coolant requires special measures to prevent external heat fluxes to the working volume of the system. This problem may be solved by using special thermal insulation.

10.4. Heat transfer in vacuum insulation

In vacuum insulation elements, heat is transferred by radiation and rarefied gas. In radiative heat exchange some quantity of internal energy of one body converts into energy of quanta of electromagnetic radiation (photons) which transfers the energy to the second body where radiation enehgy is partly absorbed, i.e. converted into internal energy of the second body.

Heat exchange in a rarefied gas essentially depends on its degree of rarefaction, characterized by the Knudsen number: **Kn = l / L,** where **l** is the mean free path length of molecules; **L** is the characteristic linear size of a system (e.g. width of gap between the walls of a vacuum cavity).

The discreteness of a gas structure starts manifesting itself at **Kn** > 0.01. In this region, a gas cannot be considered a continuous medium, and the Fourier and Navier-Stokes equations cannot be used to analyze heat exchange processes.

At **Kn** ≥ 10, transfer processes practically cease to depend on molecule collisions. This region is referred to as free-molecular. Heat transfer by gas molecules in screen-vacuum insulation occurs in the free-molecular regime.

10.5. Cooldown of cryogenic systems

The specific features of heat exchange in cryogenic systems mean that in this case consideration should be made of a number of new problems not seen at normal temperatures. Cryogenic coolants cannot exist at ambient temperature in liquid phase.

For a cryogenic system to be set into the working state, it should be preliminarily "cooled", i.e. the walls of the equipment must be cooled from ambient temperature to the cryogenic one.

Filling of volumes with cryogenic liquids is accompanied by boiling processes on the walls and by phase changes on a free liquid surface.

When the first amount of cryogenic liquid enters the volume whose wall temperature exceeds T_{cr2}, film boiling appears in the lower part of the volume. Whilst ever the walls are cooled, in this cross-section the film boiling regime is replaced by transition and nucleate boiling; when the saturation temperature T_s is reached, boiling ceases and further cooling occurs with single-phase natural convection. Meanwhile, the liquid level increases and film boiling appears in the next highest sections.

If the liquid is subcooled to a saturation temperature, then on the section filled with the liquid the heat flux removed from the wall is partly used in evaporation and partly in heating of the liquid. The heated liquid rises and forms, near the interface, a stratified layer with temperature exceeding the mean-mass temperature of the liquid.

In the upper part of the volume filled with the vapor, the heat flux is removed from the wall through natural convection and is used in vapor heating. Thus, the cooling of a wall at differrent cross-sections throughout its height occurs at different rates (maximum on transition and nucleate boiling sections; minimum in the upper cross-sections of the volume). Therefore the heat flux over the wall to the regions of intensive heat release is of significance.

The rate of increase in liquid level is determined (except for the wall temperature) by the hydraulic resistance of the inlet and drain sections and by the pressure at the volume inlet and drain outlet. The process is self-regulated to provide the liquid flow into the volume and the vapor removal through the drain valve. If evaporation on the boiling sections and on the free liquid surface exceeds the vapor removal, then the volume pressure increases. In this case, liquid in-flow decreases, and subcooling increases, resulting in less evaporation and in a higher drain vapor flow-rate. The vapor mass balance is thus recovered.

In cases where the vapor removal predominantes over evaporation, the volume pressure drops, thereby lowering the drain flow-rate and improving evaporation (i.e. this shifts the system towards the equilibrium position).

Since with change of pressure, the processes are rearranged not instantaneously but with finite (although rather high) rates, the volume pressure fluctuates (for a given filling level) around a certain value.

Questions

1. What are the distinctive features of the heat conduction processes at low temperatures?
2. What are the specific features of the hydrodynamics of two-phase flows at low temperatures?
3. What is the phisical mechanism of heat exchange on nucleate boiling?
4. How is the heat on film boiling transferred?
5. What film boiling regimes are observed for cryo-fluid channel flow?
6. What is the phisical mechanism of transient boiling heat exchange?
7. What are the specific features of vapor condensation on a cold wall?
8. What are the main features of heat transfer in vacuum insulation elements?

11

GAS LIQUEFACTION AND SEPARATION SYSTEMS

11.1. Gas Liquefaction

Cryogenic liquefaction installations are most widely used to manufacture liquid methane, oxygen, nitrogen, hydrogen, and helium. The difference in the cycles used is mainly specified by two factors: (1) essential difference in condensation temperatures of substances to be liquefied from 112 K for methane to 4.2 K for helium; (2) difference in the installation output from several liters to tens of thousands of liquid liters per hour. It is natural that larger scale installations should have more economic and usually more complex cycles.

11.1.1. Methane liquefaction

Industrial methane liquefaction is performed in connection with the problem of storing and transporting liquid natural gas whose main component is methane. The technological schemes of the units do not differ from those generally adopted in cryoengineering. The classical cascad cycle characterized by its well-known high economy is often used. Its main shortcoming is abundance of machining, as each cryoagent is compressed by an individual compressor. Therefore, a uniflow version of a cascad cycle is of great interest. As seen from Fig. 11.1, a uniflow version uses a simple compressor to compress a multicomponent mixture of gases with different boiling points (for example, propane-ethylene-methane). This mixture, circulating in a closed loop, cools and liquefies the main natural gas flowing via appropriate heat-exchanger channels.

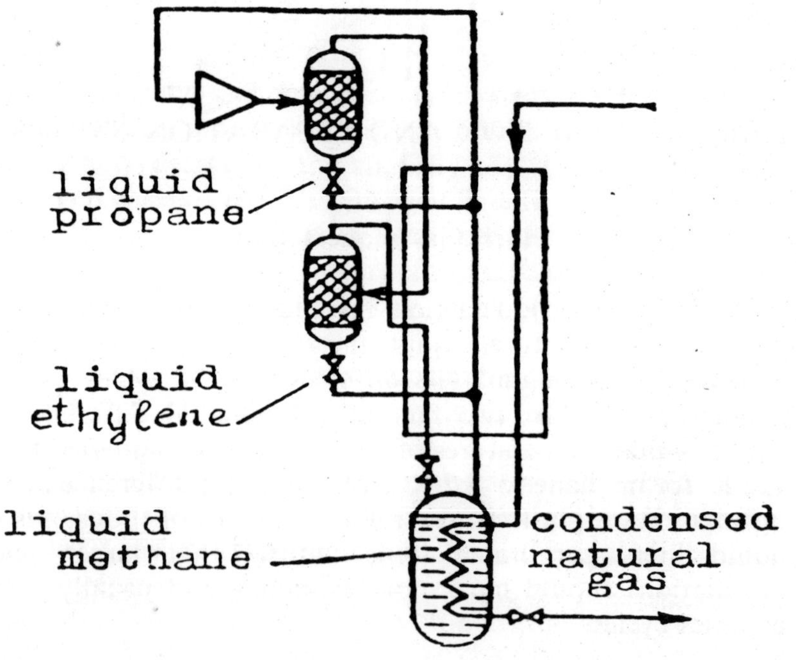

Fig. 11.1. Schematic of a single-flow cascade natural gas liquefaction cycle.

11.1.2. The natural gas liquefaction

In the natural gas liquefaction technique, the uniflow cascade cycle operated with hydrocarbon mixtures has enjoyed wide use.

Industrial natural gas liquefaction units are also known to make wide use of cycles with expansion and circulation of pure nitrogen and when mixed with methane. Sometimes, a cascade includes several expanders set at different temperature levels to increase the exergic efficiency of a cycle.

Energy consumption in large industrial methane liquefaction units (the capacity of some reaching 15000 $m \cdot h^{-1}$ can be 0.4÷0.5 $kW \cdot h \cdot kg^{-1}$ (disregarding the elevated initial pressure). The degree of liquefaction (liquefied amount to compressor feed ratio) ranges from 0.12 to 0.22.

11.1.3. Air liquefaction

On a large scale, air liquefaction is performed as an intermediate stage in air separators to produce oxygen and nitrogen.

A particular low-pressure cycle air liquefaction system including an expander and regenerator was developed by P. Kapitza.

Cryogenic gas machines (CGM) operating in the Stirling cycle are very useful for air liquefaction.

11.1.4. Hydrogen liquefaction

Nowadays, hydrogen liquefaction is realized mainly by the following three methods:

(1) throttling of preliminarily compressed and cooled hydrogen (use of the Joule-Thomson effect);

(2) expansion of compressed H_2 with absorption of external work;

(3) utilization of cold generated at a level below 25 K by a special cryogenic gas machine or in a discrete cryogenic machine (helium or neon cycle).

Throttling is a very simple method. It therefore has widespread use everywhere, including small laboratoty liquefiers (with capacity from several liters per hour to tens of liters per hour) or large pilot units with capacity of about 1000 $l \cdot h^{-1}$.

Liquid nitrogen is the commonest for preliminary cooling of compressed hydrogen below the temperature of the inverse Joule-Thomson effect. It allows temperatures of about 63 K to be obtained at low-pressure boiling.

In throttle liquefiers of low and medium capacity, a preliminary cooling temperature is kept at a level of 65 to 70 K, compression pressure being between 12 and 15 MPa.

The use of low temperature cycles with gas expansion for hydrogen liquefaction allows reduction of energy consumption by 20 ÷ 25% compared to throttle cycles. However, complication of the unit and employment of expander liquefier justifies expanding machine application only in large systems.

The combined production of large amounts of liquid hydrogen and deuterium extraction seems to be promising.

11.1.5. Hydrogen slush

Hydrogen slush, or a mixture of solid and liquid hydrogen is of interest since it promises long-term storage and filling of space craft tanks. This is because the slush density is 81.53 $kg \cdot m^{-3}$, being 70.8 $kg \cdot m^{-3}$ for liquid hydrogen. Converting hydrogen slush (50% of a solid phase) into liquid hydrogen at atmospheric pressure requires the supply of some 83 $kJ \cdot kg^{-1}$ heat. Thus, 13% more slush mass can be loaded per unit volume than liquid hydrogen. When heat is supplied, hydrogen slush will not evaporate first, as the heat will be used up in solid phase melting. This increases the total hydrogen storage terms without losses. As the hydrogen slush is at triple point conditions ($\mathbf{T}$ = 13.8 K, $\mathbf{p} \approx$ 7 kPa), in storing and transporting the above volume of slush is filled with gaseous helium to prevent air penetration. Hydrogen slush is produced by pumping out (evacuation) the vapor above the liquid (under adiabatic conditions) with solid hydrogen ice crushing and mixing. In using such a technique, about 15.4% hydrogen should be pumped out from liquid hydrogen at the normal boiling point to produce 50% slush at the triple point.

11.1.6. Neon liquefaction

Liquid neon as a cryoagent has a number of advantages over hydrogen. These are mainly the high evaporation heat per unit liquid volume (some 3.2 times higher than for hydrogen); chemical inertia and explosion safety; relatively simple liquefaction and storage.

Gaseous neon is obtained as a by-product of air separation. Neon liquefaction is realized in the same processes as hydrogen: throttling of compressed and pre-cooled gas or using two-stage cryogenic gas machines.

Small amounts of neon can be periodically liquefied using liquid hydrogen as an external cold source. To produce 1.0 *l* of liquid neon requires about 4.5 *l* of liquid hydrogen.

11.1.7. Helium liquefaction

It is common practice to perform helium liquefaction through the Joule-Thomson effect, throttling at pressures between 1.6 and 2.5 MPa. As the helium inversion temperature is low (about 35 K), the compressed gas should be pre-cooled to 10 ÷15 K. Of foreign cryoagents, only liquid hydrogen can be used for this purpose. In modern helium liquefiers, expander or, sometimes, cryogenic gas machines are frequently employed.

Dispite some inconveniences in using hydrogen (for example, explosion hazard), hydrogen-cooled helium liquefiers are still in use. They are common where application of large amounts of hydrogen is coupled with other objectives, for example, in liquid hydrogen bubble chambers. Liquid nitrogen (1 ÷ 2 *l* per 1 *l* of liquid helium) is usually used in the pre-cooling stage.

The required liquid hydrogen is poured from outside or liquefied in a separate cycle performed in a common cryoblock of a helium liquefier. In the latter case, liquid hydrogen is poured without hazard, and hydrogen cycle cold is used more effectively. The liquid hydrogen flow-rate depends on the heat exchanger efficiency, and constitutes $0.8 \div 1.2$ *l* per 1 litre of liquid helium.

The helium precooling temperature, when liquid hydrogen is used, is limited by its triple point temperature (about 14 K). At this temperature, the optimum compressed helium pressure is from 2.2 to 2.5 MPa, while the compressed helium part is about 18%. The total energy consumption in producing liquid helium pre-cooled by nitrogen and hydrogen are 3.5 to 5.0 $kW \cdot h \cdot l^{-1}$.

In 1934, P.Kapitza proposed and realized a helium liquifier using the cold, produced in a piston-type expander, rather than liquid hydrogen. In Fig. 11.2, a basic diagram of the helium liquefied developed by P. Kapitza is given. Considerable difficulties were involved in designing an expander whose piston could move in a cylinder at low temperature (about 15 K) without friction or essential expanded gas leaks. These difficulties were overcome by sealing piston slot (small clearance between the piston and cylinder).

Recently, liquefiers of this type have found wide use. In order to increase the thermodynamic efficiency of a cryogenic cycle, several expanders at different temperature levels may be used. This allows helium liquefaction without extra cryoagents.

Modern helium piston expanders are characterized by high efficiency (adiabatic efficiency is $80 \div 83\%$ expanded gas temperature berween 10 and 12 K), low leakage through a piston-cylinder slot (3 to 5% of processed gas), comparatively long life (up to 20,000 hours). Piston expanders are used in low and medium capacity units (to $50 \div 100$ $l \cdot h^{-1}$).

In high-capacity units, more than 100 $l \cdot h^{-1}$, the use of turbine-driven expanders is reasonable. Along with having good adiabatic effeciency of 75 to 80%, such expanders are highly reliable and long-lasting. The latter is due to the fact that there are no sliding friction elements working at low temperatures in turbo-expanders.

Turbo-expanders can be used in low-capacity units

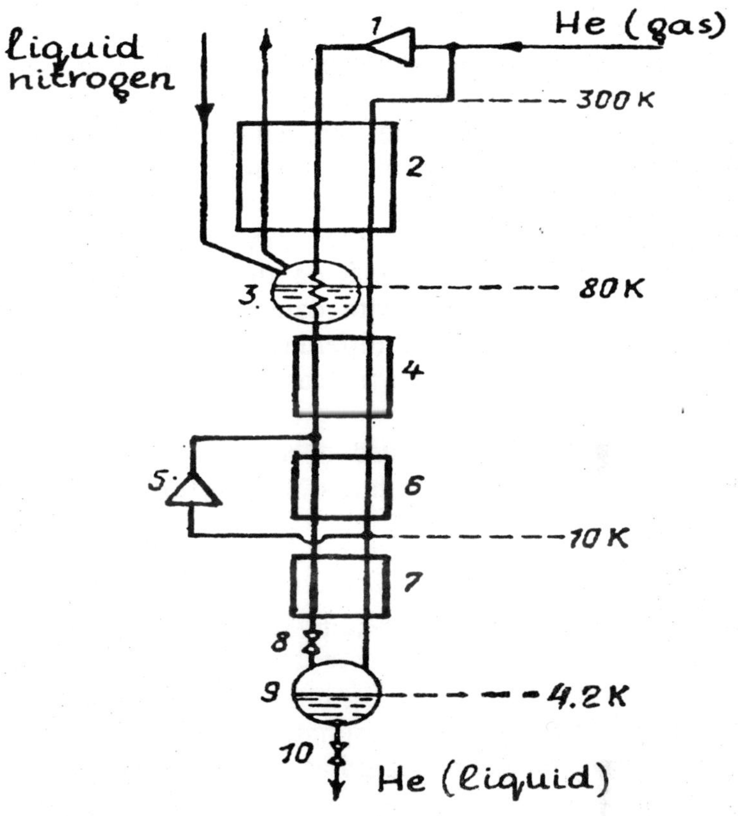

Fig. 11.2. Basic diagram of the Kapitza helium liquefier: 1, piston compressor; 2, 4, 6, 7, heat exchangers; 3, liquid nitrogen bath; 5, piston expander; 8, throttle valve; 9, liquid helium tank; 10, liquid helium discharge valve.

(some companies manufacture turbo-expander helium liquefier of 8 to 20 $l \cdot h^{-1}$ capacity) but this seems unreasonable. Such expanders are characterized by small impeller sizes (15 ÷ 20 mm in dia), high rotor rotational frequency (up to 10^4 s^{-1}), higher requirements on manufacturing accuracy. The cost of a high-rotational helium turbo-expander is much higher than a piston expander of the same capacity.

In modern expanding liquefiers, helium liquefaction constitutes 6% for two expanders without extra cryoagents; 9% for one expander with a liquid nitrogen bath; 12% for two expanders with a liquid nitrogen bath. The energy rate for helium liquefaction in such units constitutes 2 to 2.5 $kW \cdot h \cdot l^{-1}$.

Some increase in the effeciency of helium expanding liquefiers is possible by replacing the throttle valve by a piston expander operated with damp vapor. Mechanically, such an expander is little different from an ordinary gas machine. However, its usage increases helium liquefaction by 35 ÷ 40%.

Philips have developed a compact helium liquefier using two cryogenic gas machines, with throttled helium flow cooling on the cold surfaces.

11.2. Separation of gas mixtures

Separating gas mixtures into components requires a gas mixture to be cooled to a phase change temperature. In addition, a product to be separated is frequently produced as liquid.

Most of the initial gas mixture is atmospheric air. Such processes are also used to separate natural and secondary gases, as well as rare isotopes: deuterium from hydrogen, He^3 from He^4. Used cycles must be set, taking account of properties of working substance, production output, and product, and some other factors.

11.2.1 Air separation

The dry air composition at sea level is given in Table 11.1.

One- and two-column rectifiers are used for air separation.

Table 11.1. Dry air composition

Component	Volume concentration, %	Mass concentration, %
Oxygen	20.9416	23.75
Nitrogen	78.084	75.52
Argon	0.394	1.28
Carbon dioxide	0.03	0.046
Neon	1.82×10^{-3}	1.2×10^{-3}
Helium	5.24×10^{-4}	7.2×10^{-5}
Crypton	1.14×10^{-4}	3.3×10^{-4}
Xenon	8.7×10^{-6}	3.9×10^{-5}
Hydrogen	5×10^{-5}	3.5×10^{-6}
Methane	1.5×10^{-4}	0.8×10^{-4}
Nitrogen oxide	5×10^{-5}	8×10^{-5}
Ozone	$10^{-6}\div10^{-5}$	$10^{-6}\div10^{-5}$

In a one-column rectifier (Fig. 11.3), compressed air precooled in a heat exchanger is condensed in a coil set in the column cube, to then be throttled (approximately to atmospheric pressure) and supplied as a liquid to the upper rectifier tray. In a limited case, the emerging vapors are in equilibrium with liquid air, containing 21% O_2 , and cannot therefore contain less than 7% oxygen. In practice, air content is some 10%. The liquid flowing down the trays is enriched with oxygen. Therefore, any concentration oxygen can be obtained in the cube.

In a one-column unit, about 65% air-oxygen is extracted. In order to obtain 1 m^3 oxygen, 8.5 to 10 m^3 of air should be processed.

A two-column rectifier consists of a high-pressure rectifying column (botton) and a low-pressure column above it (top) with a condenser evaporator between them. In the condenser-evaporator, nitrogen vapors from the bottom column are condensed in tubes, while liquid oxygen flowing down from the top column boils in the tube space.

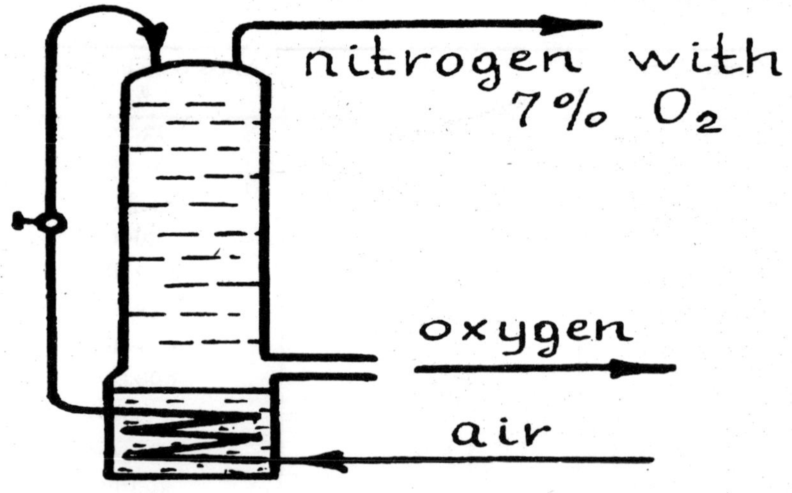

Fig. 11.3. Single-column rectifier.

Industrial air-separators include cryogenic (expander or throttle) machines along with technological units (rectifying columns, heat exchangers, adsorbers).

Inert gas production is especially economic when there are by-products in processing large amounts of air in oxygen or nitrogen production.

Inert gas production falls into two stages:

(1) production of inert gas concentrate (the so-called "raw gas");

(2) concentrate refining.

In Fig. 11.4, a basic diagram is shown for the production of the initial concentrate ("raw gas") of inert gases in an air separator.

11.2.2 Separation of hydrocarbon gas mixtures

Industrial separation of hydrocarbon gas mixtures (natural and coke gases, as well as gases of petroleum crack-

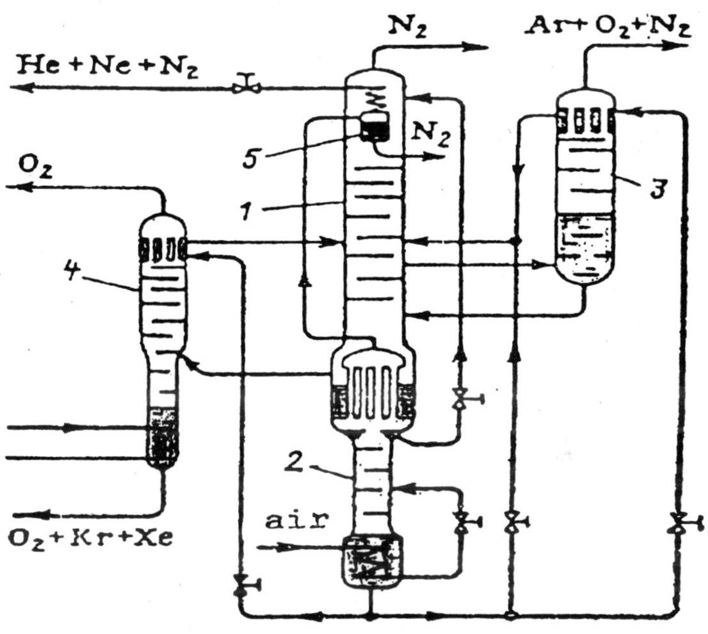

Fig. 11.4. Scheme of inert gas extraction from air: 1, upper column; 2, lower column,; 3, argon column; 4, krypton column; 5, neon-helium mixture deaerator.

ing and pyrolysis) is performed using low-temperature, absorption and adsorption techniques.

The low-temperature method is based on fractional condensation of gaseous mixture with subsequent rectification. For example, helium extraction from natural gases relies on the high difference in condensation temperatures of helium and other gas components: when a compressed natural gas is cooled, all its components are liquefied, helium remaining in gaseous state.

The absorption method uses selective dissolution of some gaseous mixture components in a certain liquid (absorbent), other components being gaseous. The absorbed gases are separated from the saturated solution by its heating in a desorption column.

The adsorption method makes use of the large difference in absorbing various gas mixture components by activated carbon or other adsorbent. In industry, hypersorption is used, or continuous adsorption on a moving sorbent bed arranged in column units.

Diffusion separation of gas mixtures through the membranes of different materials seems promising. The difficulties involved in realization of this method are explained by the low total permeability of the materials with good separation coefficient.

Questions

1. Explain the sequence of operations of methane liquefaction.
2. What are the specific features of natural gas liquefaction?
3. What methods are used for hydrogen liquefaction?
4. What are the advantages of using hydrogen sludge ice (slush)?
5. What methods are used for helium liquefaction?
6. How does rectifying equipment for air separation operate?
7. What are the main features of hydrocarbon gas mixture separation?

12

MICROCRYOGENIC SYSTEMS

12.1. Specific features of microcryogenic systems

The specific features of microcryogenic systems particularly pronounced in scientific and production aspects mean that they are a special trend of cryogenic engineering.

A small-scale model of a cryogenic system sharply increases the relative magnitude of external and internal losses. The influence of this factor is so considerable that in the majority of cases, revision of the available recommendations on choice of cooling methods, cryoparameters of plants is essential. A number of cycles and designs which have shown a good perfomance in cryogenic engineering prove to have low efficiency, and in some cases are inapplicable. On the other hand, preference is often given to cycles, machines and apparatus not used earlier, or else it is necessary to find new, more effective cryogenic cooling methods or to refine known techniques. It is not accidental that in the development of microcryogenic techniques, gas cryogenic machines based on the Stirling, or Gifford-McMagon cycles etc. are being strictly updated and finding wider use. Classical cryogenic cooling methods are being revived because of the use of throttling effect, adiabatic expansion.

Energy consumption, mass, overall dimensions, and operation time are the most important specifications of any cryogenic plant. The most stringent requirements on these indices are specified for microcryogenic systems since the latter are often used in plants with limited energy resources, mass and overall dimensions.

The higher energy load of the main elements of microcryogenic systems and a sharp increase in relative losses cause considerable complication of the problem of obtaining parameters applicable for the used. Attaining high efficiency under these conditions requires solution of a number of complex engineering problems with regard to the specific features of working processes in such plants. Most characteristic are: unsteadiness of heat and mass transfer processes occurring at variable mass, pressure and temperature; high heat load; high rate of processes; particular conditions for interaction between the environment and the object to be cooled that affect the nature of thermodynamic and thermal processes, etc.

An extremely wide range of resource indices of microcryogenic systems depending on purpose, use and type of system is typical of such systems. The required time of continuous operation varies from dozens of seconds to several thousand hours, and the service life from 1 min to 25000 h of more.

The design and operation of microcryogenic systems are strongly affected by operating conditions and other requirements which, in particular, cover maintenance of normal operation over a wide range of enviroment parameters (temperature, pressure, moisture); resistance to mechanical load (vibration, shock, acceleration); self-contained operation and simple servicing, etc.

Successful development of any microcryogenic plant and optimum realization of its parameters may be achieved only by thorough consideration of the specific features of the design of devices to be cooled and their operation. Moreover, in a number of cases, it is most rational to consider a cryostated object and a microcryogenic plant as a unit complex device, the parameters of its elements closely interrelated.

However, this statement does not exclude possible wide unification and standardization of microcryogenic work-pieces since a larger number of low-temperature devices and instruments may be included, with a known degree of convention, into one of relatively small groups of one-type work-pieces having more or less similar cryoparameters and design characteristics.

The characteristics of objects to be cooled are:

- cryostating temperature $\mathbf{T_c}$ and required accuracy to maintain it ($\Delta \mathbf{T_c}$ range or permissible rate of varying temperature $\mathbf{dT_c / d\tau}$);

- internal heat release $\mathbf{Q_i}$ and environment heat fluxes $\mathbf{Q_e}$; the sum of these heat fluxes $\mathbf{Q_\Sigma = Q_i + Q_e}$ is the useful heat load of a microcryogenic plant and corresponds to its minimum permissible refrigerating capacity under steady conditions;

- the reduced heat capacity $\mathbf{C_r}$ of elements and assemblies of a plant to be cooled which, together with $\mathbf{Q_i}$ and $\mathbf{Q_e}$, determines the required refrigerating capacity of a system under starting conditions;

- the conditions for thermal and design integration of a microcryogenic plant with an object to be cooled: distance between them, possible designs of a joint assembly, heat transfer regime, etc.

The physical basis behind operation, design and parameters of the numerous plants that need cryogenic cooling can be rather diverse. As a result, we confine ourselves to a short description of only some of the most typical devices of one of the new fields of technology, namely, cryoelectronics, in which microcryogenic work-pieces and advances in low-temperature physics are widely and successively used. In their cryoparameters, some plants to be cooled (cryomedical equipment, superconducting electric and magnetic systems, etc.) are close to various types of cryoelectronic devices.

A generalized cryogenic characteristic of cryogenic work-pieces is shown in Fig. 12.1. It is simple and, to a considerable extent, conventional in nature and does not exhaust all combinations of parameters possible in practice; additional information about some of these work-pieces, IR receivers and SHF devices, is summarized in Table 12.1 and Table 12.2.

Cryo-electronic devices may be classified by the following:

- according to cryostating temperature, these are devi-

Table 12.1. Cryoparameters of IR receivers.

Device	Spectral sensitivity range, mμ	Cryoparameters				
		T_c K	ΔT_c K	Q_e W	Q_i W	C_r J
Superconducting Sn-based bolometer	Submillimeter region	3.7	10^{-3}	0.5	10^{-3}	2000
Cd, Sn, Cu, Zn-alloyed Ge photoresistances	40	4.2	0.5	1	10^{-3}	800
Hg, Cu-alloyed Ge photoresistances	25	0÷35	1.0	1.5	10^{-3}	400
Cd Hg Te photoresistances and photodiodes	15	80	0.5	0.5	10^{-2}	200
In Sb photoresistances	6	77	0.5	0.5	10^{-3}	200
In As photodiodes and PbS photoresitances	4	80 ÷ 150	5.0	0.5	10^{-3}	200

ces with helium (up to 15 K), neon-hydrogen (up to 70 K), and nitrogen (above 70 K) cooling ranges;

- according to heat load value, these are devices with small internal heat release and relatively small mass of elements to be cooled (IR receivers), with considerable heat release but small heat capacity of elements (semiconductor

Table 12.2. Cryoparameters of SHF devices.

Device	Cryoparameters				
	T_c K	ΔT_c K	Q_e W	Q_i W	C_r J
Quantum paramagnetic amplifiers (masers)	1.5÷4.2	0.5	3	0.5	5×10^4
Parametric SHF amplifiers	4.2÷77	3	5	0.1	5×10^2
Integral SHF device	20÷85	5	10	0.5	4×10^5

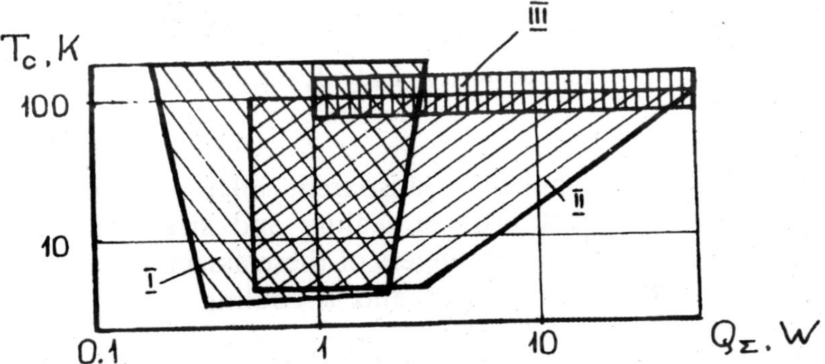

Fig. 12.1. Regions of cryostating temperatures and heat loads for various devices: I, IR receivers; II, SHF devices; III, SQG.

quantum generators - SQG), and finally with large mass and large heat capacity of elements to be cooled but small internal heat release (SHF devices).

Cryocooled IR receivers constitute the most representative group of cryoelectronic devices often used together with microcryogenic systems in various fields. A typical design of a photoreceiver (Fig. 12.2) contains, in addition to a cooling photosensitive element, a thermally insulating va-

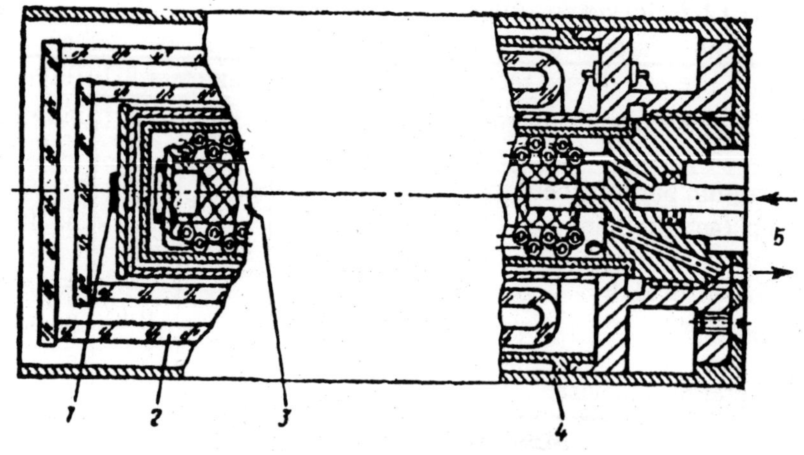

Fig. 12.2. Scheme of a cryocooled IR receiver:
1, photosensitive element; 2, vacuum flask;
3, microcooler; 4, body; 5, refrigerant.

cuum flask with optically transparent inlet windows and body assemblies that provide sealing and strength. Sizes, configuration and cryoparameters of these assemblies depend on the type of photoreceiver and a cooling system to be connected.

For example, there are cryocooled photoreceivers that have no thermal insulation, photoreceivers built in a refrigerant container or located directly on the "cold" heat of the cooling device, etc.

The main feature of photoreceivers is the low heat release in the sensitive element; only in multielement low-Ohmic semiconductor devices can the quantity $\mathbf{Q_i}$ reach $0.5 \div 1$ W. As a rule, enviroment heat fluxes $\mathbf{Q_e}$ and reduced heat capacity $\mathbf{C_r}$ are relatively low, so allowing photoreceivers to be connected with any type of microcryogenic plant. Requirements for the cryostating temperature stability are simple for a good many devices of this type (except for superconductor bolometers, and also photoreceivers in which low-frequency noise level is regulated).

Semiconductor quantum generators (SQG), i.e. lasers and light diodes, are very close in design to photoreceivers. From the cryogenic technique viewpoint, the main difference is that considerable heat power, up to $10 \div 20$ W, per channel is released on an active element of SQG usually mounted in the same zone of the device as the photoreceivers photosensitive element. This SQG distinctive feature requires heat transfer conditions to be accurately estimated in the active element zone and sets a limit on the choice of possible variants of cooling systems.

Cryo-electronic SHF devices, are, by design, an independent category of low-temperature equipment, first and foremost, distinguished by relatively large values of $\mathbf{Q}_e$ and $\mathbf{C}_r$; this is connected with the large size of refrigerated SHF unit-equipped cryostats, noticeable heat fluxes through wave-guides and other facilities for a SHF signal input, as well as with a refrigerated assembly mass much greater than for IR receivers and SQGs. In many cases, the design of this type of device does not allow microcryogenic plant to be mounted in the close vicinity of an object to be cooled, and prevents local cooling of an active element (e.g. diode in a parametric amplifier), thereby causing additional cold losses.

Recently, most frequently use is made of combined cooling of the entire amplification system: inlet SHF filter, parametric diode and other units; in this case, the entire multifunction device is located in a cryostat, whose inner container surface is cooled to working temperature.

The design of a refrigerated device and the physical characteristics of its active (sensitive) element are also responsible for the way in which the device is linked to the microcryogenic plant; assembly of their connection must meet the requirements that provide normal operation of a thermostated device under certain parameters of a microcryogenic plant:

- vibrations of a plant, if there are any, must have minimum transfer to the object to be cooled;
- temperature drop and cold losses in the joint zone

must be minimum; in many cases, temperature fluctuations that appear on the cold heat of a microcryogenic plant may make the device characteristics and must be avoided by taking special design measures;

- the location of both the microcryogenic plant and the device to be cooled, as well as the design of the assembly to be connected to them should be chosen from the conditions under which the device is screened from the influence of electromagnetic fields arising during plant operation. It is also important to provide the sealing and strength of a joint assembly in order to exclude possible damage of a refrigerated object during cooling system operation.

12.2. Main quality indices of microcryogenic systems

Depending on the purposes and problems of the comparative analysis, the quality and degree of thermodynamic perfection may be evaluated by the set of all the characteristics, by separate index groups, or by determi-ning single criteria. In practice, the following indices are most significant: cryostating temperature; useful and reduced refrigerating capacity; exergetic efficiency; start time of an operating regime; relative mass of a system; relative volume of a system.

The useful refrigerating capacity Q_2 means the maximum amount of heat to be removed by a system per unit time at cryostating temperature T_c.

The reduced refrigerating capacity Q_r corresponds to the minimum power necessary to remove heat flux Q_2 to the enviroment and coincides with the energy of this heat flux. The magnitude Q_r is determined from

$$Q_r = Q_2 \tau_e = Q_2 \frac{T_0 - T_c}{T_c},$$

where τ_e is the exergetic temperature function and T_0 is the environment temperature.

For microcryogenic systems, $\mathbf{Q_r}$ does not exceed 500 W.

The exergetic efficiency of closed cryogenic cycle plants is the reduced ratio of refrigerating capacity to supplied exergetic power:

$$\eta_e = \frac{\mathbf{Q_r}}{\mathbf{N_c}} .$$

For electric energy systems, the supplied exergetic power coincides with the electric power **N**, while for heat supply systems it coincides with the power of the working part of the heat flux.

The exergetic efficiency of stored compressed or liquefied gas systems is equal to the ratio of the real exergetic result of processes actually occurring in them, i.e. the exergy of the total heat from an object to be cooled to the maximum possible result to be obtained, assuming reversible process.

The exergetic efficiency is a universal index for characterizing microcryogenic systems, irrespective of their type and the temperature of operating. It also shows the degree of approaching a thermodynamically ideal process of production of cryotemperatures. As a result, use of the exergetic efficiency as an estimate criterion for the thermodynamic perfection of microcryogenic systems takes on high significance in practice since it gives an indication of the achieved level and resources of the efficiency increase.

In practice, the concept of a relative power $\mathbf{n = N / Q_2}$ is often used for comparing energy indices of electric energy systems. The corresponding exergetic index (relative reduced power)

$$\mathbf{n}_e = \frac{\mathbf{N}}{\mathbf{Q_r}} = \frac{1}{\eta_e} .$$

Similarly, relative reduced mass and overall characteristics

$$\mathbf{m}_r = \frac{\mathbf{M}}{\mathbf{Q}_r}; \qquad \mathbf{v}_r = \frac{\mathbf{V}}{\mathbf{Q}_r},$$

are determined; here **M** and **V** are the mass and volume of a system.

12.3. Classification of microcryogenic systems

A wide field of application, varoius different requirements and operating conditions, as well as different potentialities of the known cryogenic cooling methods when applied to miniaturizing cryogenic plants have meant the advent of a wide nomenclature of microcryogenic systems of different types and designs. At present, there is no generally accepted classification of microcryogenic systems. This is because of the insufficient development of the basic methods of organizing and classifing cryogenic plants, and because of the great number of indicators, according to which microcryogenic systems may and must be classified. The classification proposed below covers specific features of the most important microcryogenic systems in practice, and has its main use for choosing a particular type of system. The classification is based on the following main features: basic cycle or cooling method, cryostating temperature, refrigerating capacity.

Depending on the cycle or cooling method used, microcryogenic systems are subdivided into:

- radiative cooling systems (based on radiant heat exchange with the environment);
- systems using the throttling effect of real gases;
- systems using evaporation and sublimation heat of cryogenic substances;
- systems using gas cryogenic internal hear generation machines operating in Stirling, Vulemier, Solway (Gifford-McMagon) cycles and their modifications;
- Claude and Brighton cycle systems equipped with

piston machines and turbomachines;

- electronic refrigerant systems;
- combined systems using several cooling methods.

Radiative cooling systems apply to passive ones as they operate without consuming energy; throttle plants with a compressed gas to be stored as energy source, as well as vaporization heat or sublimation systems (i.e. liquid- and solid-state) correspond to open methods; the remaining types of electronic cooler systems are called closed.

Depending on cryostating temperature, a large number of microcryogenic systems may be placed in one of four groups:

- systems operating at a nitrogen level of 80 K;
- systems operating at a neon level of 30 K;
- systems operating at a hydrogen level of 20 K;
- systems operating at a helium level of 4.5 K.

It should be noted that along with plants classifiable in the above classical thermostating levels, microcryogenic systems with higher cooling temperatures (100 K and above) are beginning to find wide use. These plants can be classified as a separate group.

It is rather difficult to classify systems by value of useful refrigerating capacity: it can simply be noted that the useful refrigerating capacity of the known microcryogenic systems is close to one of the following values: 0.5; 1; 2; 3; 5; 10; 20; 50; 50 W. Larger values are typical of nitrogen cooling level systems while smaller ones are typical of helium, hydrogen and microcryogenic helium level systems does not usually exceed 2.5 W.

12.4. Recommendations on choosing type of microcryogenic system

Use of a particular type of microcryogenic system in each specific case is determined by a set of indices and requirements, whose fulfillment provides both normall operation of an object to be cooled and a modern technological level of the system and refrigerated apparatus as a whole. However, there are always one or two indices that depend on the given conditions. These are cryostating temperature, refrigerating capacity, exergetic efficiency (or energy consumption), mass and operating time of a system. The first two indices are the main ones for all types of microcryogenic systems; the significance of the others depends on the particular conditions of their use.

The choice of microcryogenic plant type starts by determining cooling method and system scheme optimum for the given conditions. In this case, account must be taken not only of the limiting potentialities of the cooling method from the viewpoint of obtaining required temperatures, refrigerating capacity and efficiency, but also of the advantages, disadvantages and distinctive features typical of miniature plants, where underestimation may lead to incorrect conclusions and recommendations.

In particular, the following should be noted.

(1) Progressive growth of relative losses by friction, displaced heat transfer, and other losses due to scating.

(2) Increase in heat regeneration magnitude due to decrease in useful refrigerating capacity of a system. The ultimately high efficiency of heat exchangers, in which heat is being regenerated, is the main condition that not only determines the system perfection but also, in some cases, the principal way of designing it.

(3) Increasing the influence of design factors on the efficiency end parameters of the cooling cycle. With decreasing machine sizes, the relative "dead" volume of compression and expansion cavities increass, heat fluxes through design elements of low-temperature assemblies and relative losses by the hydraulic resistance of channels and pipelines, etc., increases. This exerts an influence on the nature and efficiency of working processes and, hence, on the indices of real microcryogenic systems.

(4) Important influence of the quality of the insulation on the parameters of many microcryogenic systems to be used to cool stored liquid or liquefied refrigerant, e.g. the

insulation quality shows that a system with acceptable characteristics may be designed.

Taking the mentioned specific features into account, the possible use of the known cycles and cryogenic cooling methods in microcryogenic systems can be considered.

Fig. 12.3 slows a plot of the exergetic efficiency versus cryostating temperature for some cycles and schemes of cryogenic plants. The curves show that for each cooling method a relatively narrow temperature range exists, above which a systems high efficiency is maintained. Gas cryogenic Stirling cycle machines (curves 5, 6) have maximum efficiency over a temperature range 70 ÷150 K, while throttle systems in which nitrogen serves as refrigerant (curve 3), have minimum efficiency. Throttle gas-mixture systems (curve 8) and those equipped with regenerative turbo-machines (curve 4) occupy an intermediate position. At lower temperatures (70 to 4.2 K), the absolute value of the efficiency of all types of systems drops sharply. However, the advantage of the efficiency is retained for gas cryogenic machine and expander systems (curve 7). Utilization of vapor compression plants (curve 1) and thermoelectric coolers (curve 2) in microcryogenic systems is affected by temperature conditions.

The above data apply to rather large-scale plants and therefore cannot completely reflect energy potentialities of the considered cycles and methods in miniature systems. Moreover, some new cycles widely used in modern microcryogenics (Vulemier, Gifford-McMagon, combined cycles, etc.) have been omitted.

Throttle cryogenic systems are the most simplest to miniaturize, but have the lowest efficiency when compared with other types of closed systems. This results from the throttling imperfection and low isothermal efficiency of microcompressors. Means of improving the energy characteristics of high-pressure microcompressors are limited. However, the efficiency of throttle microcryogenic systems may be elevated considerably by improving the thermo-physical properties of the refrigerants.

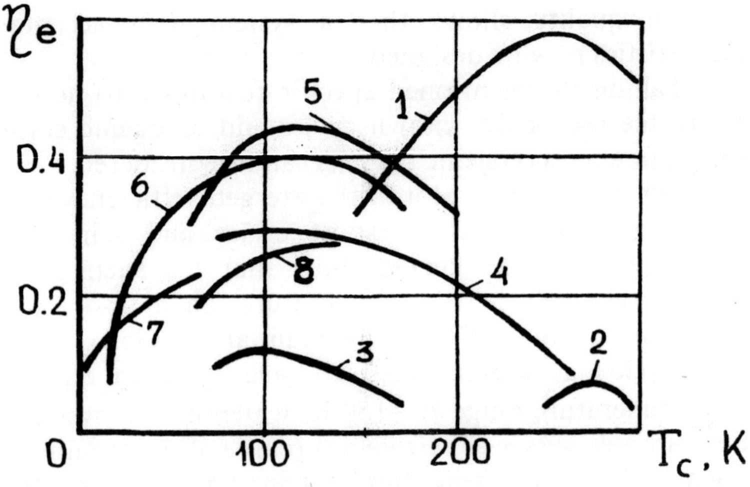

Fig. 12.3. Exergetic efficiency η_e of low-temperature plants as a function of cryostating temperature T_c: 1, vapor compression plants; 2, thermoelectric coolers; 3, throttle nitrogen systems; 4, systems with regenerative turbo-machines; 5, 6, gas Stirling cycle machines; 7, expander systems; 8, throttle gas-mixture systems.

In miniaturizing, Stirling cycle systems retain the highest efficiency compared to other cycle and cooling methods. Modern modifications of the two- and three-stage expansion Stirling machines enable a cryostating temperature of 9 ÷ 12 K to be obtained, thereby retaining, in this case, rather a high system efficiency.

Recently, Vulemier cycle machines have found wider use in microcryogenic engineering. These machines are not equipped with a mechanical compressor; a heat energy source is used to produce the cold. The efficiency of the initial thermodynamic Vulemier cycle is equal to that of the Stirling cycle while the efficiency of real Vulemier cycle systems is somewhat lower, a fact connected with design difficulties. The lower temperature limit of Vulemier cycle systems amounts to approx. 10 K.

In 1959, Gifford and McMagon proposed an original cryogenic plant scheme, according to which non-equilibrium expansion of a working body occurs, with energy transferred to the environment in the form of heat. The exergetic efficiency of micocryogenic heat pump cycle systems is approx. 1.5÷4.0 times lower than that of Stirling cycle machines, since cooling uses the irreversible exhaust process. However, on lowering a cryostating temperature, the efficiency difference is noticeably decreased, and at a temperature of 4.2 K it is only 20÷30% for helium throttle cascade systems. In connection with this, heat pump cycle machines have found their wider use at temperatures 3.5 to 20 K.

Turbo- and piston expander Claude and Brighton cycle systems may, in principle, have rather a high exergetic efficiency, especially at helium temperature level. However, in practice there are, at present, major production difficulties associated with miniaturizing expanders. Even the most modern versions of microcryogenic plants with these cycles have efficiency no higher than that of heat pump systems.

Analysis of the characteristics and potentialities of existing plants, as well as of the tendencies to developing cycles, schemes and cooling methods, enables a number of general recommendations to be made on regards choosing type of microcryogenic system.

Fig. 12.4 and Fig. 12.5 illustrate fields of application for various types of microcryogenic systems versus required cryostating temperature, refrigerating capacity and service life. The use of microcryogenic systems versus refrigerating capacity and service life shown in these figures is affected by the energy and overall dimension characteristics of each type of system optimum for a given cryostating temperature.

Passive radiative systems are promising for objects to be cryostated at temperatures above 80 K and long service life. In this case, it should be remembered that as temperature increases, the refrigerating capacity of radiative systems may increase, without varying their overall dimensions and mass.

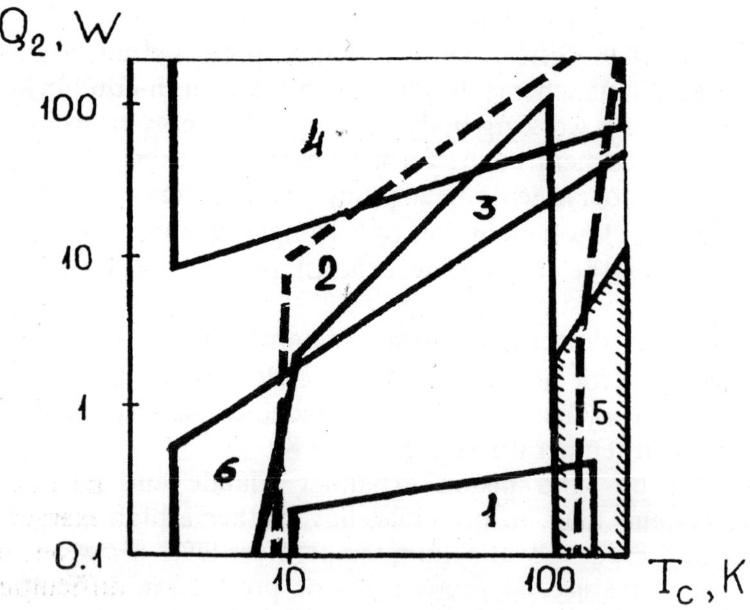

Fig. 12.4. Fields of application of microcryogenic systems (vs refrigerating capacity and temperature); 1, solid refrigerant systems; 2, Solway (Gifford-McMagon) cycle systems; 3, Striling and Vulemier cycle systems; 4, systems equipped turbo-machines and rotating piston machines; 5, radiative systems; 6, closed throttle systems.

The choice of open-system type optimum for service life and refrigerating capacity is limited by the need for acceptable mass and overall dimensions. At a refrigerating capacity up to 3 W, the optimum service life of open throttle systems does not excees 15 h, and of liquid systems 100 h (Fig. 12.5).

It is advisable to use solid refrigerant cooling systems with a heat load up to 0.5 W over a temperature range 10 ÷ 110 K. They may provide a service life of 100 ÷ 10000 h. Raising the refrigerating capacity of systems sharply increases

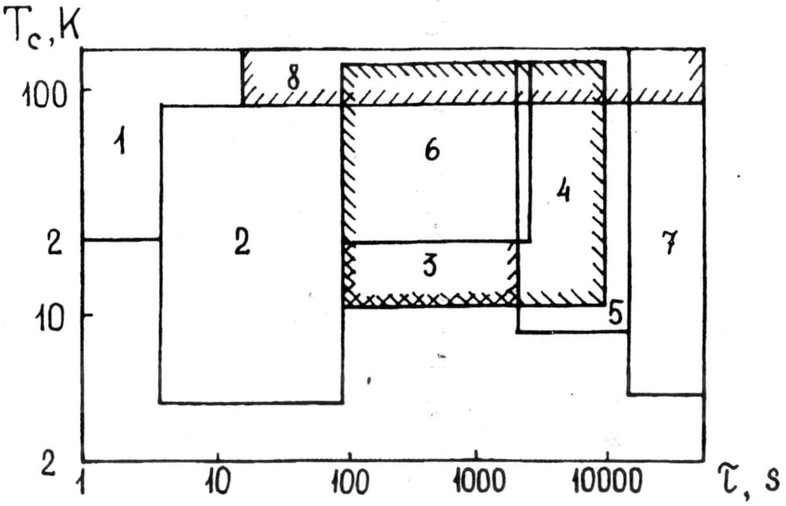

Fig. 12.5. Fields of application of microcryogenic systems (vs service life and temperature): 1, open throttle systems; 2, liquid refrigerant systems; 3, Stirling and Vulemier cycle systems; 4, solid refrigetant systems; 5, Solway (Gifford-McMagon) systems; 6, closed throttle systems; 7, systems equipped with turbo-machines and rotating piston machines; 8, radiative systems.

the mass and overall dimensions for a given service life, and renders them unsuitable for use when compared with other types of systems.

Specification of suitable fields of application of closed microcryogenic systems is a more intricate problem. On the whole, the known types of closed systems can be compared by cryostating temperature and refrigerating capacity, and the advantage of the exergetic efficiency is not always a determining factor while choosing an optimal type of plant, since under real conditions account must be taken of the entire set of different factors and, above all, the service life.

Closed throttle systems may be utilized over the whole cryogenic temperature range 4 to 120 K at a refrigerating capacity from fractions of a watt to 10 ÷ 15 W; however, the range 30 ÷ 100 K is most promising. The service life of these systems amounts to 1.5 ÷ 2 thousand hours and may be increased to 2.5 ÷ 3 thousand hours when lubrication compressors are used. Despite the low thermodynamic efficiency, closed throttle systems have found wide use in microcryogenic engineering. Use of such systems is determined in some cases by specific needs (e.g. the final cascade, helium microrefrigerators), and in other cases by advantages of their design, mainly by their simple connection to the object to be cooled and convenient location in the apparatus.

Vulemier and also Stirling cycle systems have the same aim of applicability in terms of temperature and refrigerating capacity. In terms of efficiency, at temperatures below 20 K, Vulemier systems have noticeable disadvantages in relation to Stirling machices, though in terms of a service life and vibration characteristics they very often compare favourably. The service life of real Vulemier cycle systems amounts to 1500 ÷ 3000 h and can be increased to 4000 ÷ 8000 h. One drawback of Stirling and Vulemier cycle machines is the need to carry out periodic precautionary maintenance, replacing refrigerant contaminated by seal-wear products.

It is profitable to use thermal pump cycle systems when an elevated service life (10 thousand hours or more) is needed. A minimum temperature at which effective cooling is provided by a thermal pump cycle system amounts to 7 K or less. At temperatures below 15 K, the heat pump efficiency drops less sharply than with Stirling cycle machines, and the efficiency difference of these systems becomes much less marked. Service life advantages of a heat pump are especially notable. Advantages of such plants also involve the low oscillations of the low-temperature assembly and the possibility for locating it remote from the compressor unit. An optimal field of a application of microcryogenic heat pump cycle systems covers stationary plants with a service time over 5000 h and a cryostating temperature below 18 K.

Systems equipped with turbo-machines and rotating piston machines are used for cooling objects at a temperature above 4 K and a refrigerating capacity more than 2 W. On increase of refrigerating capacity, the characteristics of such systems improve more rapidly than those of other plants. In particular, this applies to turbo-machive systems. The advantages of the systems under consideration imply the possibility of reaching almost infinite service life (more than 200 thousand hours).

The search for an optimal type of a closed microcryogenic system cannot be restricted just to analysing the main potentialities of different cycle and schemes. Account should also be taken of the achived technological level of developing each type of system. It is therefore of interest to compare real microcryogenic systems in terms of the two most important criteria: exergetic efficiency η_e and relative reduced mass $\mathbf{m}_r$.

Fig. 12.6 and Fig. 12.7 show plots of the efficiency and relative mass as a function of reduced refrigerating capacity $\mathbf{Q}_r$ for specimens of microcryogenic systems over a temperature range 3.5 ÷100 K.

Use of reduced (exergetic) refrigerating capacity $\mathbf{Q}_r$ as a criterion enables characteristics of micro-cryogenic systems to be compared irrespective of their temperature level.

For convenience, systems are subdivied into four groups in terms of cryostating temperature Tc: 70 ÷ 100, 25 ÷ 35, 10 ÷ 20 and 3.5 ÷ 4.5 K.

Analysing the data in Fig. 12.5 and Fig. 12.7, the following may be noted. Real systems even of one type for the same value of $\mathbf{Q}_r$ have different efficiency and relative mass. This is attributed to different levels of developments and to the effect of a number of particular factors (operation conditions, purpose, etc.) on characteristics of specific plants.

Energy and mass indices of systems as a function of cooling method, refrigerating capacity and temperature level

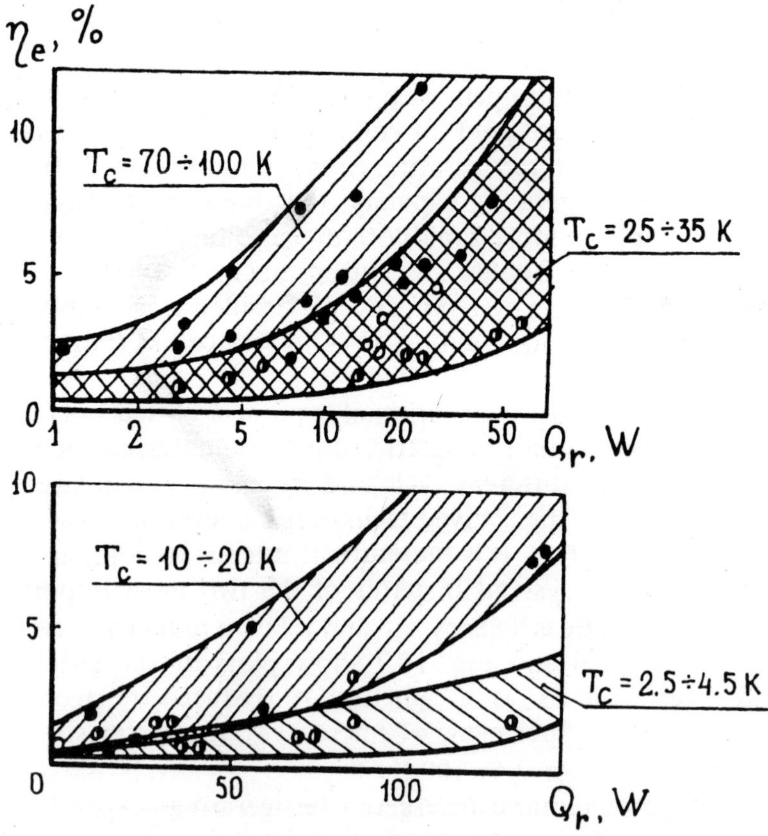

Fig. 12.6. Efficiency indices of microcryogenic systems: o - closed throttle systems; ◑ - Solway (Gifford-McMagon) cycle systems; ● - Stirling and Vulemier cycle systems.

are sufficiently clear and support the above recommendations for selecting a plant type.

With increase of reduced refrigerating capacity and cryostating temperature, the efficiency increases and the relative mass of all types of systems decreases. The boundaries of the hatched zones in Fig. 12.6 and Fig. 12.7 characterize the modern technological level of systems developed for each of the four temperature groups. In Fig. 12.6, the

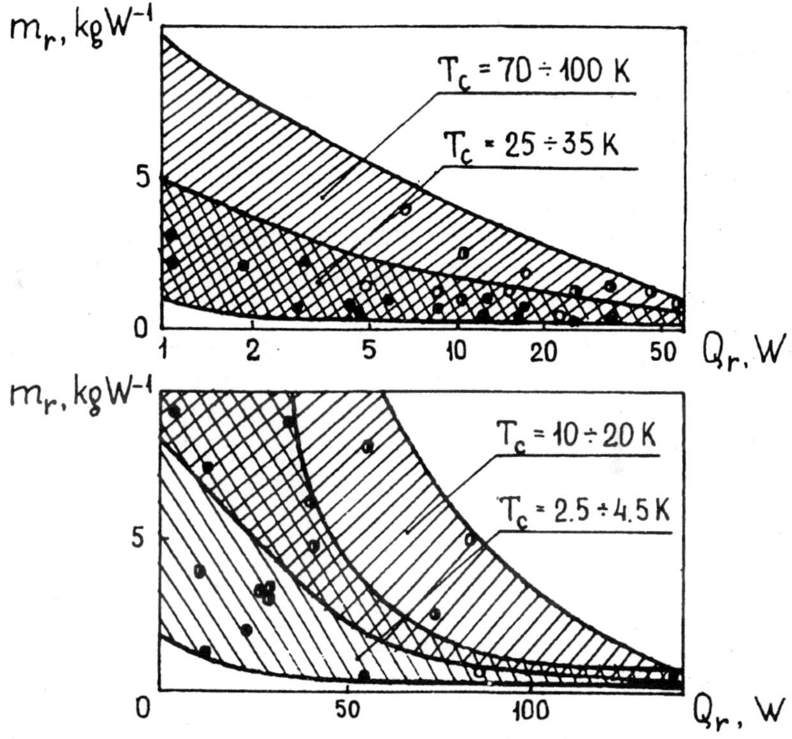

Fig. 12.7. Specific mass indices of microcryogenic systems: o - closed throttle systems; ◐ - Solway (Gifford-McMagon) cycle systems; ● - Stirling and Vulemier cycle systems.

upper boundaries of the zones correspond to the efficiency indices of the best of the existing systems. Stirling and Vulemier cycle systems have the nighest efficiency (Fig. 12.6a). At $\mathbf{Q_r} = 1 \div 25$ W, η_e for such plants reaches $10 \div 11\%$ for systems with $\mathbf{T_c} = 70 \div 100$ K and $6 \div 8\%$ for systems with $\mathbf{T_c} = 25 \div 35$ K. The lower boundary corresponds to η_e for throttle plants and heat pump cycle systems. Of a similar nature is η_e versus $\mathbf{Q_r}$ for systems with $\mathbf{T_c} = 3.5 \div 20$ K (Fig. 12.6b) but the cryostating temperature and refrigerating capacity effect on the efficiency is more clearly pronounced here. It is characteristic that at very low temperatures and small $\mathbf{Q_r}$, the efficiency difference of Stirling and heat pump cycle systems is much smaller than that for systems with larger $\mathbf{T_c}$ and $\mathbf{Q_r}$. This is attributed to the increasing effect of the above factors manifesting itself as cryogenic system are minictriced.

The effect of the refrigerating capacity $\mathbf{Q_r}$ on specific mass microcryogenic system indices $\mathbf{m_r}$ is shown in Fig. 12.7. For all temperature levels and values of refrigerating capacity, Stirling and Vulemier cycle systems have the smallest relative mass compared with other types of plants. Throttle systems and Gifford-McMagon cycle systems have the largest values of $\mathbf{m_r}$. The $\mathbf{m_r}$ range for systems with $\mathbf{T_c} = 3.5 \div 20$ K is much wider than for systems with higher cooling level. This is especially seen for small-refrigerating capacity systems and is mainly explained by the stronger influence of the miniaturization factor.

Questions

1. What difficulties arise in designing cryogenics systems in relation to their miniaturization?
2. What are the main characteristics of objects to be called using microcryogenic systems?
3. What are the basic quality indices of microcryogenic systems?
4. How does the exergetic efficiency decrease on reduction of the refrigerating capacity of different-type systems?
5. How does the specific mass of microcryogenic systems increase on decrease of a systems refrigerating capacity?

13

LIQUID AND SOLID-STATE MICROCRYOGENIC SYSTEMS

13.1. Features of the liquid and solid-state microcryogenic systems

Cryogenic cooling devices exploiting the latent heat of evaporation or sublimation of refrigerants have long been used in research and commercial plants. These are simple in design, reliable and convenient to operate, consume minimum energy and, in many cases, compare favourably with other types of cooling systems in terms of the following factors: mass, overall dimensions, process start up time, cooling temperature stability, etc. As a result of these advantages, despite the development of a great number of different types of adequately improved microcryogenic plants, liquid and solid-state cooling systems, as before, remain an important and widely used group of microcryogenic equipment.

In the majority of cases, the operation of such devices includes two main regimes: storage of cryogenic fluid (or solid refrigerant) in a special thermally insulated container at low temperature and evaporation of refrigerant, with the evaporation (or sublimation) heat used to remove the heat from the object to be cooled. Some designs utilize a gas obtained due to a phase change; in this case, an object is cooled either due to rising gas temperature or due to its subsequent throttling, accompanied by cold regeneration in an ordinary throttle microcooler.

Thus, in the general case, a microcryogenic system operating on stored liquid or solid refrigerant contains a thermally insulated container, a receiver that provides the connection with the object to be cooled, pipelines or refrigerant pipes connecting these main assemblies of a system, a device for filling a system with a refrigerant, as well as monitor and control units.

As refrigerants of liquid and solid-state cooling systems, use is made mainly of ofdinary cryogenic substances. Thermophysical parameters of some of them necessary to develop and calculate cooling systems are listed in Table 13.1.

Table 13.1. Characteristics of various refrigerants.

Refrigerant	Boiling point, K, at a pressure of 0.1 MPa	Specific latent vaporization heat, $kJ \cdot kg^{-1}$	Fluid density $kg \cdot m^{-3}$	Triple point temperature, K	Specific melting heat, $J \cdot kg^{-1}$	Solid state density, $kg \cdot m^{-3}$
Helium	4.22	20.43	125	-	4.56	160
Hydrogen	20.38	441.70	71	13.95	58.20	90
Neon	27.10	86.25	1206	24.54	16.62	440
Nitrogen	77.36	198.45	804	63.15	25.75	950
Argon	87.29	162.03	393	83.81	29.52	1700
Methane	111.67	510.79	426	90.66	58.60	520
Ethane	184.53	489.86	546	89.82	95.46	-
Carbon dioxide	194.70*	571.08	1180	216.55	199.04	1600
Ammonia	239.76	1369	682	165.42	332.01	800

* Sublimation temperature

From Table 13.1 it is seen that the choice of optimal refrigerant depends on the cooling temperature range and specific use of a plant. For the range $30 \div 40$ K, for example, liquid neon is highly promising, its high density favoring a reduction in container size and heat flux value. Moreover, neon temperature depends only weakly on pressure, thereby stabilizing the cooling temperature by means of a very simple pressure regulator. Solid methane, ethane and ammonia, having a higher latent heat of sublimation than argon or nitrogen, are promising refrigerants for temperatures above 80 K.

The possibility or advisability of using liquid and solid-state systems and the choice of refrigerant are specified by the operating conditions of an object to be cooled; liquid systems are more commonly utilized in system with limited preliminary storage period and relatively short working cycle time, while solid-state ones, are used where there is limited energy consumption under long continuous vacuum operating conditions.

So for, generally accepted criteria for the perfection of liquid and solid-state systems have not yet been laid down. As a result, comparison of different designs and the search for optimal solutions are somewhat difficult. Use of exergetic characteristics appears to be inadvisable since in the majority of cases, the available liquefied or solid refrigerant energy is almost completely used up in compensating heat fluxes and is dispersed into the environment. In other words, the exergetic efficiency of liquid and solid-state systems intended for cooling small internal heat release objects is close to zero.

For liquid cooling systems, a suitable criterion might possibly be the so-called relative time characteristic

$$\tau = \frac{r}{\left(1 + \frac{M_c + M_i}{M_l}\right) Q} \frac{T_0 - T_l}{T_0},$$

where T_0, T_l are ambient and refrigerant temperatures, respectively; r is the latent specific evaporation heat of refrigerant; M_l, M_c and M_i are the masses of liquid refrigerant, container and thermal insulation.

Use of this criterion allows comparison of different systems of stored liquid cryogenic substance (in any case, long storage) and estimation of valid ways of improving such systems.

13.2. Liquid cooling-systems

Three main designs of systems of this type are used:

- combined: refrigerant container is, in design, connected to the object to be cooled;

- remote: refrigerant is transferred from a container to an object to be cooled via a special pipeline;

- refrigerant evaporation remote: refrigerant is transferred in the form of a compressed gas to the throttle micro-cooler inlet.

A liquid cooling system with remote refrigerant supply usually contains a thermally insulated container, a receiver, refrigerant pipes, a device for filling a system with a refrigerant, monitor and control units. To elevate reservoir pressure, an electric heater, a special coil-evaporator, or a heat valve may be used, or else a compressed gas is supplied from an external source.

The following main variants of liquid systems of this type are designed and used: continuous refrigerant supply, pulse flow, program fluid supply from a reservoir to an object to be cooled. In the latter two varieties, refrigerant is supplied automatically, a cutting-off device switching on through a probe signal in the presence of fluid in the cooled-object zone; a program variant allowing for heat load oscillation due to varying ambient temperature enables the refrigerating capacity to be regulated more exactly according to the heat load value and provides the most economical operating conditions.

As regards design, the developed liquid cooling systems with remote refrigerant transfer are very diverse and limited by service conditions. As an example, Fig. 13.1 shows a cryosurgical device "Cryolor" (refrigerating capacity is up to 3 W; working cycle, 5 min; liquid nitrogen container volume, 15 cm^3; total mass, about 0.3 kg).

Plant operating with a double-change of phase state of a working substance are a variety if liquid cooling systems. Refrigerant is stored in the liquid state in reservoirs of such plants; then at the beginning of the working cycle the reservoir pressure is increased owing to refrigerant evaporation by electrical heater up to 10 ÷ 20 MPa, and high pressure gas is supplied via pipeline to a throttle microcooler connec-ted to the object to be cooled.

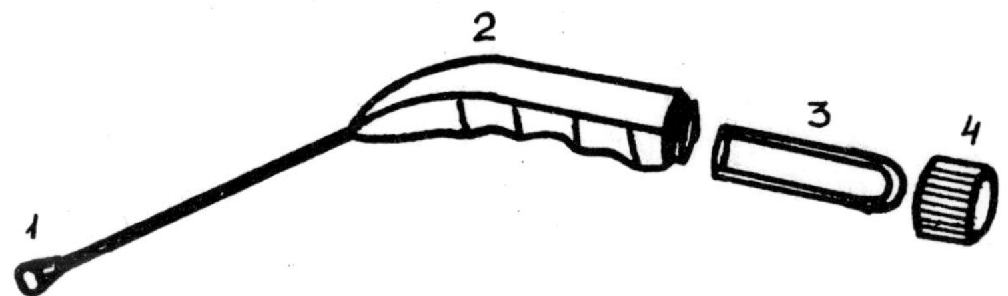

Fig. 13.1. Cruosurgical device "Cryolor":
1, detachable cryoinstrument; 2, pipe handle; 3, vacuum container; 4, seal.

Advantages of such a plant are: relative volume of refrigerant storage reservoir and high degree of purity of the working gas to be supplied to the cooler; the main drawback is the complex reservoir design required.

A liquid system is calculated and developed by optimizing its overall dimensions and mass characteristics, which are described by the following equations:

$$M_l = \Delta M + \frac{2\lambda_i \pi D_1 D_2 \Delta T}{r(D_2 - D_1)}\tau + \frac{Q_p}{r}\tau ; \qquad (13.1)$$

$$M_c = \pi D_1^2 l \rho_c ; \qquad (13.2)$$

$$M_i = \frac{\pi}{6}(D_2^3 - D_1^3)\rho_i . \qquad (13.3)$$

Here ΔM is the useful fluid mass; λ_i the insulation thermal conductivity; D_1, D_2 the inner and outer diameters of the spherical container housing; ΔT the temperature difference between fluid and ambient medium; Q_p the heat flux via a pipe; τ the time (fluid storage or transfer conditions); l the container wall thickness; ρ_c and ρ_i the material density of the container body and insulation, respectively.

Eqn. (13.1) determines the fluid mass as a sum of the amount of fluid used for cooling an object and fluid evaporation losses.

Eqn. (13.2) determines the container mass as a function of container sizes and material density of the container body. Often, other design elements (presence of supports, couplings, etc.) are taken into account in the form of an additional mass M_l. The insulation mass is determined by eqn. (13.3).

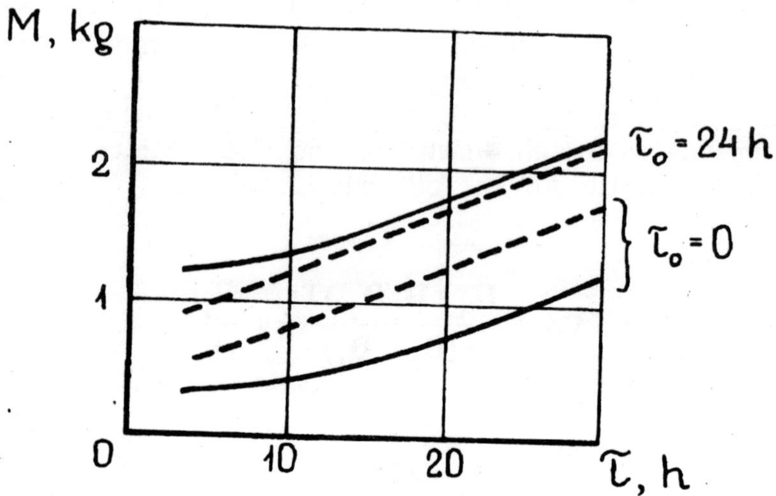

Fig. 13.2. Liquid plant mass vs operation time for different times of preliminary storage τ_0 (solid lines, combined version; dashed lines, remote system).

These equations should be naturally used with regard to a real cooling system scheme; an optimal solution corresponds to a minimum in the sum of $\mathbf{M}_c$, $\mathbf{M}_l$ and $\mathbf{M}_i$ at given τ. For different liquid plants, the mass and sizes of plants are found as a function of operation time; for nitrogen systems these data are plotted in Fig. 13.2.

13.3. Solid-state cooling systems

Over the last $10 \div 15$ years a number of developments and studies have been made of a new type of microcryogenic systems: stored solid cryogenic substance plants. The advent of such plants has resulted from some of their advantages, especially manifest in space equipment:

- smaller mass compared with liquid systems because of the higher values of specific latent heat of phase changes (sublimation heat is higher than the boiling heat for the same refrigerant) and the larger density of the solid refrigerant;
- no problem on phase separation under zero-gravity conditions.

Solid-state systems used to cool small inner heat release objects (IR receivers) allow continuous cryogening to be provided for 10^4 h or more, under certain conditions.

The main elements of a solids of a solid-state micro-cryogenic plant are a thermally insulated liquefied refrigerant container, a device to remove refrigerant vapor and to keep container pressure constant, and a refrigerant pipe connected to the object to be cooled.

The overall dimensions and mass characteristics of such systems depend on the service life, thermal insulation quality and refrigerant parameters; the heat releases of the device to be cooled are usually small compared with total heat fluxes and exert little influence on a plants size and mass.

The choice of refrigerant is the main factor in determining the design and characteristics of a plant. Table 13.2

slows working temperature ranges and upper limit of vapor pressures (lower limit 0.01 kPa) widely used in solid-state refrigerant systems.

Table 13.2. Characteristics of various refrigerants.

	T, K	p, kPa
Hydrogen	8.3 ÷ 14	7.5
Neon	13.5 ÷ 24	32
Oxygen	48.0 ÷ 54	0.13
Nitrogen	43.4 ÷ 63	12.7
Argon	47.8 ÷ 83	66.5
Methane	59.8 ÷ 90	10.7
Ethane	95.0 ÷ 104	-
Carbon dioxide	125.0 ÷ 194	100.0

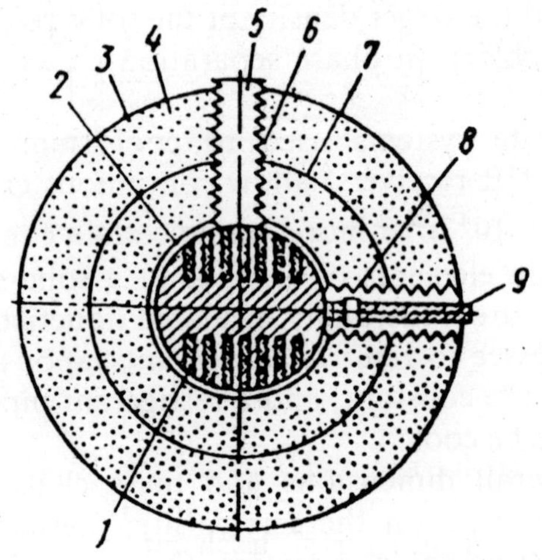

Fig. 13.3. Shceme of a one-cascade solid-state plant (USA): 1, solid refrigerant; 2, inner container; 3, outer container; 4, vacuum insulationl 5, vapor removal device; 6, 8, siphons; 7, screen; 9, refrigerant pipe.

Keeping smaller constant container pressures brings with it considerable difficulties. The highest working temperature of a solid-state plant is determined by the triple point. Other refrigerant characteristics necessary to calculate solid-state plants were cited in Table 13.1.

Plant designs can be subdivided into one- and two-cascade types (Fig. 13.3 and Fig. 13.4). Moreover, so-called combined systems are known in which a throttle cycle is used in combination with solid cryogenic substance storage.

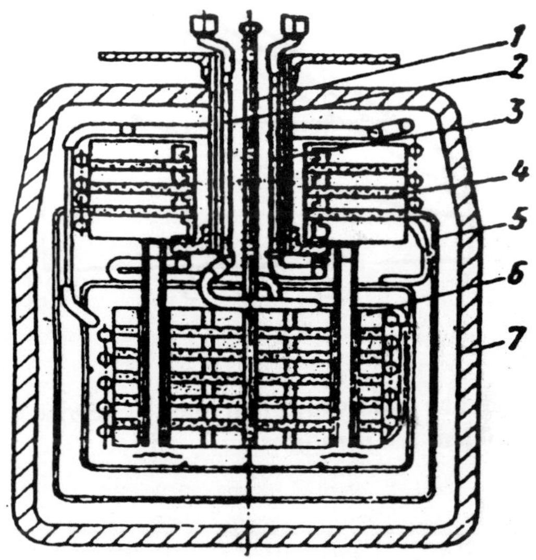

Fig. 13.4. Two-cascade solid-state plant (USA):
1, copper refrigerant pipe; 2, argon waste line; 3, carbon dioxide waste line; 4, solid carbon dioxide; 5, screen; 6, solid argon; 7, insulation.

One-cascade one-refrigerant solid-state plants are simple in design. Fig. 13.5 and Fig. 13.6 show characteristics of such plants for solid methane and hydrogen systems under the following conditions: internal heat release of the object to be cooled is 100 mW, ambient temperature is 300 K, the container has the form of a cylinder with height equal to its diameter, the external housing mass is not allowed for (under real conditions housing may be absent).

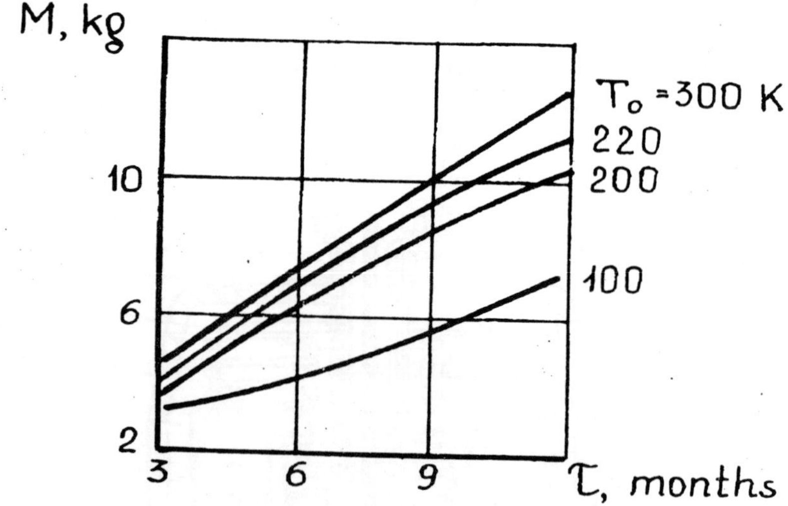

Fig. 13.5. Methane solid-state plant mass vs service life for varions ambient temperatures.

A substantial gain in a plant mass be obtained when two-cascade solid-state systems are used. A firm in the US has developed and tested an experimental specimen of a two-cascade carbon dioxide and argon plant (see Fig. 13.4); at a working cooling temperature of 50 K, about 18 mW heat release of a refrigerated IR receiver and a 1 year service life, the plant mass is 13,5 kg; in this case, 10.1 kg accounts for refrigerants (the argon mass is 6.2 kg).

The designing of a combined periodic throttle-sublimation plant is promising. The plant includes a solid refrigerant cryostat, high-pressure gas (refrigerant) container,

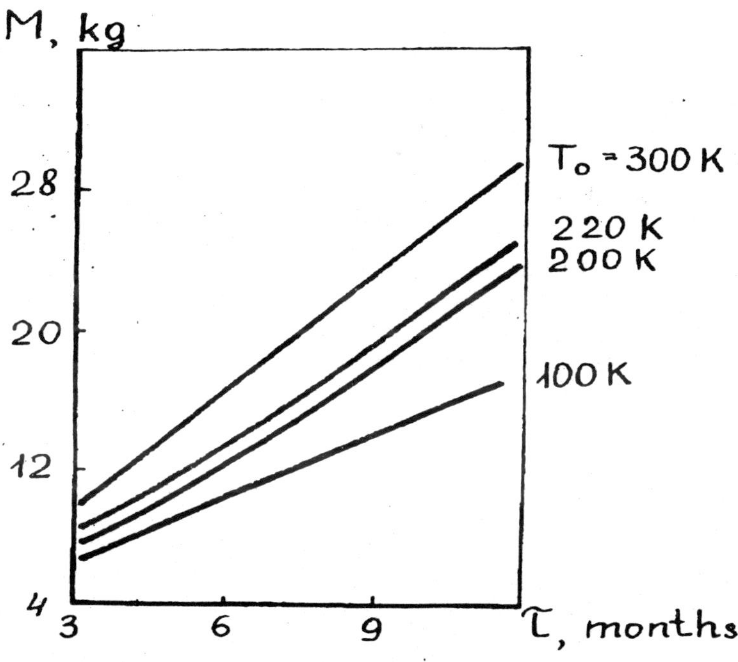

Fig. 13.6. Hydrogen solid-state plant mass vs service life.

microcooler (throttle heat exchanger), and monitor elements. The throttle cascade efficiency is increased owing to solid refrigerant cooling of a working gas. A design estimate is made for the conditions solid refrigerant: ammonia; working gas: air; heat release power of the object to be cooled 0.1 W; thermal insulation: vacuum-multilayer $\lambda_i = 3\times10^{-4}$ W·(m·K)$^{-1}$. Estimates have shown that for total service life τ of more than 3500 h such systems are advantageous over ordinary sublimation types if $\tau_w/\tau < 10\%$ (where τ_w is the total duration of a plants working cycles).

Questions

1. What are the advantages of liquid and solid-state microcryogenic systems?

2. What are the basic schemes of liquid cooling systems?

3. How are the overall dimension-mass characteristics of a liquid microcryogenic system calculated?

4. What are the advantages of solid-state cooling systems under space conditions? What cryoagents are effective for the temperature ranges 30 to 40 K and 80 to 100 K?

14

STORAGE AND TRANSPORTATION OF CRYOGENIC FLUIDS

14.1. Storage of cryogenic liquids

14.1.1. Storage vessels

Cryogenic fluid storage vessels are designed to accumulate, store and dispatch liquid cryogenic produces to the user. In certain cases, some required fluid temperature is produced in vessels in which there is also equipment to be cooled, usually by helium; such vessels are a special class, called cryostats. Their design differs markedly from that of generally accepted industrial vessels, since it is closely dependent on the overall dimensions and shape of objects to be cooled and sometimes has the possibility of mounting or removing quickly from a volume.

Depending on the designated purpose, cryogenic storage vessels may be subdivided into the following groups: stationary vessels used as a part of cryogenic systems; transport vessels to supply users with cryogenic products; laboratory vessels, whose volume and mass allow them to be moved by hand.

The vessel shape is chosen taking into account designated purpose, ease of manufacture, transport, and operation. Additionally, a vessel shape is influenced by the desire to decrease heat fluxes to the fluid to be stored.

To provide minimum heat fluxes, preference is given to a spherical vessel shape since for a sphere the surface-to-volume ratio S/V is minimum, compared to other geometries. However, in the case of spherical vessel shape, the working areas and volumes are not used effectively enough because each type-size requires new attachments to be manufactured; with increasing volume (over $5 \div 10$ m^3) vessel transportation becomes difficult. For this reason, as well as spherical vessels, there is wide use of vertical and horizontal cylindrical vessels, whose shape takes better account of the overall dimensions of transport facilities, allows more rational utilization of operating areas, and allows, because of changing a shell thickness, the manufacture of a number of vessels of different volumes. Cryogenic fluid storage vessels have a shell (usually sealed) and the space between them serves for thermal protection of the inner vessel, often using vacuum insulation. The shell shape is identical to that of the inner vessel.

The basic design of cryogenic storage vessel mainly pivots round solving the task of how to fasten the inner vessel to the shell. Usually, the inner vessel is fixed to the shell using suspensions or supports. Their design assumes a fluid-filled vessel mass and, for transport vessels, additional loads resulting from accelerating and decelerating a transport facility. Moreover, they must take up forces associated with temperature deformations of the inner vessel. At the same time, supports and suspensions are the most important elements of thermal protection.

Fig. 14.1 shows the main designs of cryogenic vessels intended to store a large number of cryogenic fluids except helium. For relatively small vessels in the shape of a sphere or a short vertical cylinder, the suspension is arranged using a central tube along the vertical axis (Fig. 14.1a), which at the same time serves for filling, draining fluid, and removing vapor: this proves to be one of the best design choices. The space between the vessel and shell is used for powder or vacuum layer insulation. This design of fastening the inner vessel to the shell is typical of almost all cryogenic storage vessels with volumes up to several hundred liters.

The thermal bridge (mouth) heat flux in a given design is minimum since the mouth length may be sufficiently long; the temperature drop at the ends of the mouth is decreased through cooling on open vessel storage with gas discharge, or else from a temperature spread in the gap phase on storage with no vapor discharge. For storage vessel transportation, the inner vessel has radial displacemant-limiting devices.

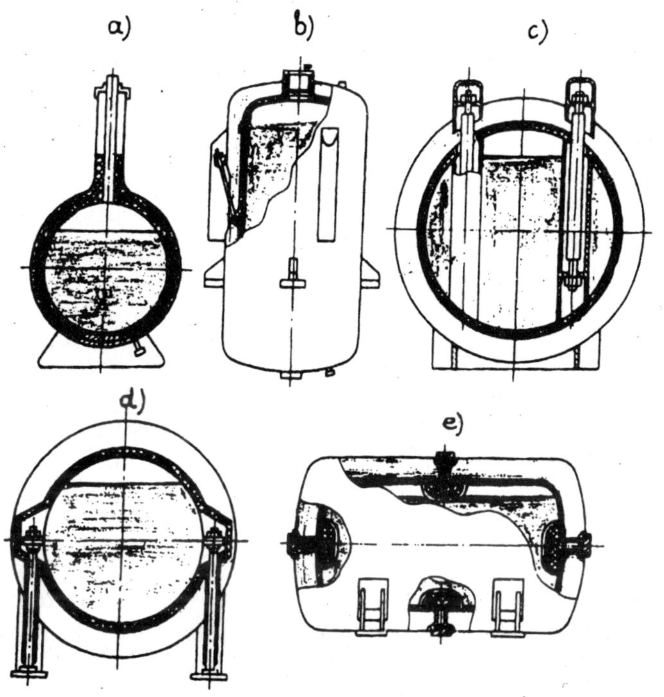

Fig. 14.1. Design scheme of cryogenic fluid storage vessels: a, fastening of the inner vessel on the central thermal bridge; b, c, suspension fastening; d, support fastening; e, plastic support fastening.

The design for larger spherical, and also horizontal and vertical cylindrical vessels allows for fastening them relative to the shell by means of suspensions (Fig. 14.1b). The location of suspensions tangential to the inner vessel enables them to be long enough to decrease the heat flux. To avoid the action of bending moments of suspension, couplings are hinged with the shell and inner vessel.

In a number of cases, for the suspensions to be lengthened, these are located in special pipes passing through the fluid vessel, the pipe cavity being connected with the insulation vessel cavity (Fig. 14.1c).

To decrease the metal consumption in a vessel, recent designs fasten the inner vessel via rigid tube supports (Fig. 14.1d). Vessel and shell supports are located co-axially and the space between them is connected with the vessel space, thus being integral with the vacuum cavity. An integral system of inner vessel and shell supports has enabled the vessel weight to be reduced.

Vertical tube supports are used in designs of large vertical and horizontal cylindrical and spherical vessels. In some vesse designs (mainly for transport), the inner vessel is fastened relative to the shell via several glass plastic supports (Fig. 14.1e). Such a design is simple in manufacturing and allows considerable load although it is less efficient as regards heat.

Very large storage vessels (more than a thousand cubic meters in volume) used for relatively high-temperature products oxygen, nitrogen, and methane are often manufactured with gas-filled insulation, giving wider choice of design. Two types of storage vessels have found wide use in practice. According to the design scheme in Fig. 14.2a, storage is arranged using an assembly of one-wall standard cylindrical vessels which when ready for operation are transported to the mounting area. Vessels are assembled in one single shell on a frame construction, their number determining the storage volume. Vertical vessel arrangement in one shell reduces the mounting area. Storage vessels of this type up to 5000 m^3 in volume, composed of standard vessels 500 m^3 and 1000 m^3 in volume, with perlite insulation, are manufactured, for example, by Nippon Sanso (Japan). The simple design and cheap cost recommend such storage arrangements for use in factories producing air separation products.

Design schemes in Fig. 14.2b,c show another type of large cryogenic product storage arrangement: these are two-wall, relatively short, vertical cylinder, flat-bottom vessels. The inner fluid vessel rests on a foundation of foamglass, i.e. a material having relatively high strength and low heat conduction. The flat bottom of the inner vessel serves only to

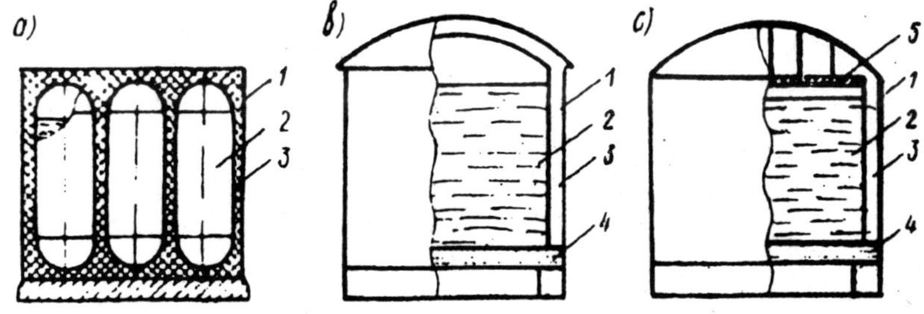

Fig. 14.2. Desing schemes of large cryogenic gas-filled insulation storage vessels: a, assembled in one shell; b, with self-supporting dome; c, with suspended roof; 1, shell; 2, innner vessel; 3, powder insulation; 4, foamglass insulation; 5, suspended roof.

provide the volume sealing and does not suffer particular stress. As a result, it may be fabricated from a relatively thin sheet. A flat-bottom vessel design has a vapor space pressure close to atmospheric, i.e. usually within $0.103 \div 0.13$ MPa.

The space between outer and inner cylinder walls is filled with powder. The vessel dome is shaped either as an ellipsoid or segment of a sphere. Two versions of such vessels are known: with self-supporting dome and with suspended roof. In designs with a self-supporting dome, (Fig. 14.2b), the vessel is a two-wall construction, thereby requiring constant control of pressure in the vapor space of the vessel and insulation cavity in order to keep them equal and so avoid crushing of the inner vessel. Vessel operation is simplified by using a suspended roof (Fig. 14.2c). A flat roof made of thermally insulating blocks is suspended by ropes from the shell dome. It has no sealing and is intended to prevent gas convection and to decrease the outer dome heat flux.

In vessels with suspended roof, the pressure in the

vessel vapor cavity is always equal to that in the insulation space; also, such a construction consumes less metal.

In addition to the difference in the dome design, flat-bottom cylindrical vessels are subdivided into thick-wall and membrane types, depending on the design of the side-cylinder walls. Design schemes in Fig. 14.2b,c include thick-wall vessels as the inner vessel can with stand the fluid pressure and column weight. The distinctive feature of membrane constructions is that the inner shell is made of a thin membrane which must possess only sealing. Fluid weight and pressure forces are transferred via a solid, sufficiently stable insulation, e.g. foamglass, to the outer supporting well. As well as steel, reinforced concrete also serves as the outer wall material. Such vessels are less metal-consuming; however, technological difficulties in providing the membrane construction sealing means that they are still less reliable.

For cryogenic vessels, ways to design output assemblies of tubes to fill and empty vessels, as well as pulse pipes, e.g. to measure fluid level by differential filled-system transducers, are important. Incorrect orientation of tubes in the space means that they flood with fluid. Thus heat fluxes grow on storage of cryogenic products and level meter indications are distorted. All pipelines via which the fluid is discharged from the vessel must be oriented in the space so the during storage these remain filled with vapor phase alone. If tubes are put out through the vessel top, this condition is satisfied automatically. If tubes are extended from the vessel bottom, these must necessarily have a lift section to form a gas valve (Fig. 14.3a). In addition the bottom tubes may be equipped with special caps. However, with the lift section absent, such caps (Fig. 14.3b) are not always reliable as gas valves.

Commercial cryogenic fluid storage and transport vessels usually use manometric level indicators, whose operation principle is based on measuring the pressure drop due to a hydrostatic fluid column.

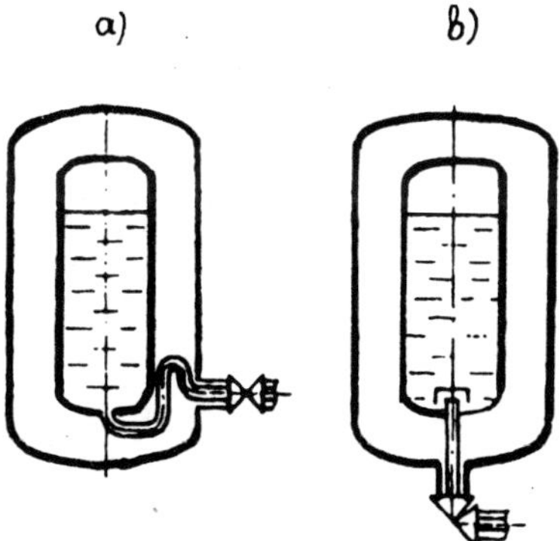

Fig. 14.3. Gas valves with bottom output of fluid tubes: a, remote gas valve; b, gas valve in the vessel.

14.1.2. Thermal protection of a product

Thermal protection of a product from environmental heat flux is arranged by using an effective thermal insulation of the vessel inner surface, a suitable design for fastening the vessel to the shell, and a scheme for binding the inner vessel with tubes. The relative fluid evaporation rate per unit time (η) sometimes called fluid evaporability, is an indicator of the efficiency of a vessels thermal protection. Here we have:

$$\eta = \left| \frac{1}{M} \frac{dM}{d\tau} \right| = \frac{Q}{V \rho_l \, r \, (1 - \varphi)}, \qquad (14.1)$$

where M is the fluid mass, $M = V \rho_l (1 - \varphi)$; V is the vessel volume; φ is the vapor void fraction; ρ_l is the fluid density; Q is the environmental heat flux; r is the specific evaporation heat.

Relative evaporation rate (evaporability) is measured and calculated for maximum filling degree of a vessel $(1 - \varphi) = 0.85 \div 0.95$. The fluid evaporability is the most important performance characteristic of a vessel, showing the ability of a given design to operate effectively as a part of a particular cryogenic system. Very high evaporability corresponds to large losses of fluid being stored and makes operating difficult, with the need for frequent refilling, rapid increase in pressure and temperature on storage with close gas discharge. The value of the daily fluid evaporation η below 1% is usually a satisfactory level for thermal protection. Since the environmental heat flux may be presented in terms of its density over the vessel surface, $\mathbf{Q = q\ S}$, and the vessel volume in terms of the product of the surface area and characteristic size $\mathbf{V = S\ l}$ (to a sphere $\mathbf{l = D/6}$), then the relative fluid evaporation rate is equal to:

$$\eta = \frac{\mathbf{q}}{\mathbf{l}\rho_l \mathbf{r}\,(\mathbf{1} - \varphi)} . \qquad (14.2)$$

This expression implies that if for a given vessel design the measured or calculated nitrogen evaporation rate is conventionally taken as unity, i.e. $\eta_{nitr} = 1$, then for oxygen $\eta_{oxyg} = 0.65$; for argon, $\eta_{arg} = 0.7$; for methane, $\eta_{meth} = 0.8$; for hydrogen, $\eta_{hydr} = 5.1$; for helium, $\eta_{hel} = 62$. Hence, it follows that requirements for thermal protection of cryogenic vessel depend strongly on thermophysical fluid properties. To obtain approximately the same evaporation rate for such fluids having different thermophysical properties, such as nitrogen, oxygen, and methane on the one hand, and for helium on the other hand, substantially different levels of thermal protection of vessels must be provided; intermediate requirements for thermal protection quality appear on hydrogen storage. Within certain limits, increasing vessel volume may decrease fluid losses or compensate insulation-quality deterioration, since the evaporation rate is inversely proportional to a characteristic linear dimension, i.e. $\eta \sim \mathbf{l}^{-1}$.

The designs of a number of helium vessels are shown in Fig. 14.4. Vessels up to several hundred liters are usually fastened to the shell by an extended mouth.

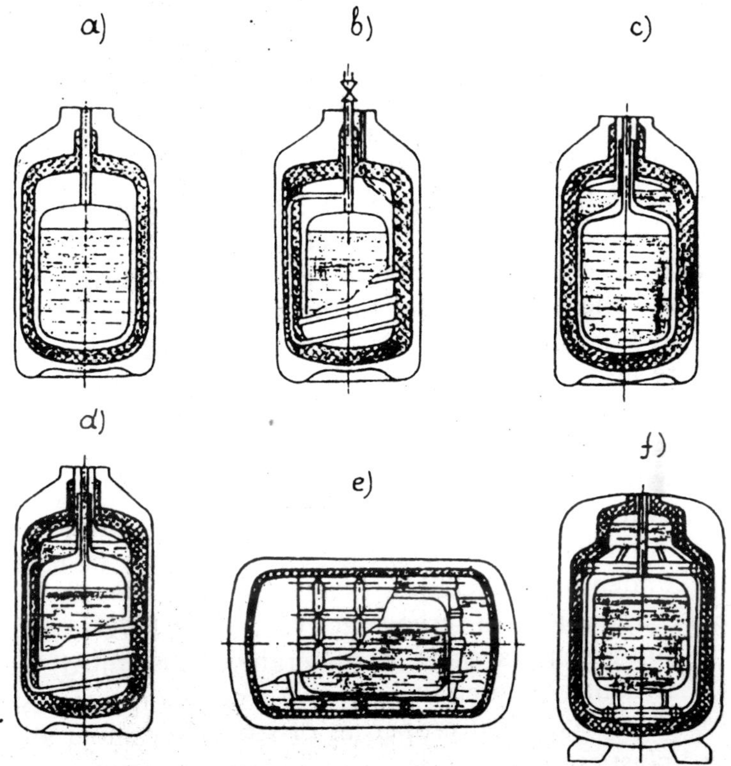

Fig. 14.4. Designs of helium reservoirs: a, b, with removed vapor-cooled screens; c, d, e, f, with nitrogen screens.

In designs using removed vapor cold, the screen is made in a high heat-conduction material (usually copper) sheet welded to the mouth (Fig. 14.4a). In some cases, for better cooling of the screen, a sufficiently long tube through which forming vapors are exhausted into the atmosphere or a gas holider (Fig. 14.4b) is welded to the screen.

Designs equipped with a nitrogen screen are more diverse. In relatively small vessels, up to several hundred liters, the nitrogen screen also takes the form of a copper sheet but welded to a nitrogen vessel located in the top part of a vacuum cavity (Fig. 14.4c). To produce lower temperature over the entire screen surface, a pipe may be welded to the sheet and natural nitrogen circulation (Fig. 14.4d) results in this pipe owing to the environmental heat flux. In larger vessels, a nitrogen screen contains one or two volumes for storing liquid nitrogen and a system of liquid nitrogen pipes located around the helium vessel. A heat-conducting metal sheet (Fig. 14.4e, f) is welded together to the pipes. A nitrogen screen is fastened by suspensions and supports directly to the shell, and the helium vessel to the nitrogen screen. This reduces heat fluxes to helium via suspensions and supports. Schemes are also known in which a removed helium vapor cooled screen is used along with a nitrogen screen.

Large stationary and transport vessels are often equipped with facilities for reverse condensation which liquefy an evaporating cryoproduct and return it as fluid to the vessel. In particular, cryogenic gas machines may be used as reverse condensation facilities.

Safe operation of cryogenic vessels and of other cryogenic equipment is quite considerably determined by choice of structural materials. At cryogenic temperatures, changes in the mechanical properties of structural materials, metals and plastics, occur. In this case, the yield point σ_y and ultimate strength σ_{str} increase, simultaneously the plasticity and shock viscosity decrease. The magnitude of the shock viscosity as a function of chemical composition, heat treatment, and other factors is the main quantitative index for a material to be suitable for operation at low temperatures. An embrittlement phenomenon, i.e. a sharp decrease in the shock viscosity at low temperatures, makes it impossible to use ordinary carbon steels. In addition to high strength at cryogenic temperatures, a metal must have an adequate safety margin to avoid damage during multiple cycles of cooling and heating in combination with loading.

The relative elongation of sample δ is a quantitative plasticity index which determines process properties of a material and its ability to re-distribute stresses in the zones of its concentrations. In practice, a magnitude $\delta > 15\%$ is satisfactory, in some cases $\delta > 10\%$. Also, a metal must possess a time structural stability independent of temperature and stress level. In view of the wide use of vacuum types of insulation, metals and welded joints require vacuum density. Metals and materials used to manufacture thermal briges must possibly have lower values of the index λ / σ_{str}, i.e. must combine high strength and low thermal conductivity λ. In designs subject to periodic heating, the total heat capacity characterized by an the index $\rho c / \sigma_{str}$ must be minimal (ρ is density, **c** is specific heat).

14.1.3. Cryoproduct storage vessels on board space ships

Under space-flight conlitions the problem on storing cryoproducts in a liquid phase is very much complicated by the zero-gravity condition, sign-variable accelerations that appear on space-craft maneuvers in orbit, large dynamic loads along the active flight trajectory, and for other reasons.

With no gravitation, the fluid-vapor interface in a vessel becomes unstable, and the vapor can displace the fluid in the gas discharge line, thereby rapidly emptying the vessel. The same effect also arises when even small accelerations appear directed along the cryoproduct vapor discharge line.

Therefore, liquid cryoproduct storage vessels on board space ships must be equipped with a special facility called a gas separator that provides free cryoproduct vapor discharge from the vessel and, at the same time, prevents fluid penetration into the discharge line.

To separate liquid and vapor phases and to maintain the liquid phase in a vessel, there are several methods based on the difference in physical properties of fluid and vapor of a cryoproduct:

(1) mechanical methods based on the density difference of vapor and fluid usually use a rotating element which transfers centrifugal forces of a two-phase mass, thus separating it and allowing vapor discharge alone;

(2) dielectrophoretic methods based on different polarization of phases in a non-uniform electric field; if a two-phase mixture is pumped for example through electrostatic high-voltage condensers, then fluid will move up preferably in a high field strength;

(3) methods based on surface tension forces to control the fluid-vapor interface;

(4) methods based on using heat exchange-refrigeration systems, in which a two-phase mixture discharged from a vessel is throttled to a lower pressure and temperature and then runs through a heat exchanger for evaporation of fluid present in the mixture.

To separate oxygen phases, use may also be made of the difference in paramagnetic properties of fluid and gas.

Designs of the earliest liquid helium storage vessels on board Earth satellites used the method based on heat exchange-refrigeration systems.

The diagram of a liquid helium storage vessel on board a space-craft is shown in Fig. 14.5. A liquid helium bath is connected with the discharge line via a phase separator which is a heat exchanger equipped with a throttle at its inlet end. Because of the hydraulic throttle resistance, the vessel pressure $\mathbf{p}_1$ and, hence, helium temperature $\mathbf{T}_1$ will be somewhat higher than pressure $\mathbf{p}_2$ and temperature $\mathbf{T}_2$ behind the phase separator, throttle. Therefore, if two-phase mixture enters the separator, then liquid phase will evaporate in the heat exchanger and only vapor will enter the vessel inlet.

Thermal insulation of space ship vessels is provided by vacuum multilayer, with cryoproduct vapor-cooled screens.

When storing liquid helium in a superfluid state (He-II), porous plugs (metal, ceramics) with very narrow slits may be used as an effective phase separator.

Specific difficulties arise when storing liquid fluorine which is the most active oxidant of all known elements. Special precautions must be taken against igniting pipelines, valves, etc. Fluorine is highly toxic, and the corresponding protective accessories are needed to work with it. Vessels used to store and transport liquid fluorine have a protective bath filled with liquid nitrogen preventing fluorine evaporation. Vessels are composed of three shells: an inner shell containing fluorine; a middle one, liquid nitrogen, and the space between the outer shell and nitrogen bath which is filled with insulation (perlite or aerogel under vacuum). Monel, stainless steel or aluminum are recommended as materials for fluorine containers. A protective film of fluorine compounds is applied to metal by special processing.

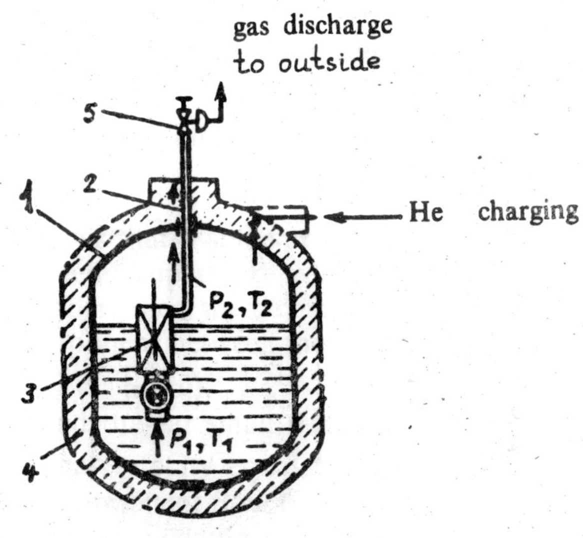

Fig. 14.5. Diagram of a liquid helium storage vessel on board a space-craft: 1, helium bath; 2, gas discharge line; 3, phase separator; 4, thermal insulation; 5, pressure regulator.

A liquid fluorine container is always closed, opened only when filling or discharging; therefore, the protactive bath must always be full of liquid nitrogen during storage and transportation.

14.2. Transporting liquid products

Transporting liquid products via cryogenic pipelines is carried out either by displacing them from vessels by gas supercharging or by pumping. In both cases, on flow discharge a certain pressure, as a rule higher than saturated vapor pressure, must by maintained above fluid surface. Raising the pressure in the vessel vapor-space increases the proposed pressure drop and fluid subcooling to a saturation temperature. With a liquid product discharged by displacing, raised pressure must compensate hydraulic losses in the pipelines; with a product discharged by pumps, raising the pressure provides work without cavitation. Usually, gasified working products are used as the supercharging gas in commercial systems.

In analysing the discharge of vessels, of most practical interest is the determination of a liquid product temperature on supercharging gas flow-rate to maintain the required pressure in the vessel vapor-space.

In discharging cryogenic products, interrelated processes occur in vessels. These are gas mixing, heat and mass transfer between the fluid surface and walls, formation and flow-down of a condensate film; downward fluid surface motion; fluid mixing.

Supercharging gas flow-rate and liquid product heating are strongly affected by gas condensation on the fluid surface. A dynamic gas-jet effect on fluid and forming a hopper above a fluid-intake connecting pipe favor this phenomenon and render it practically incontrollable. Therefore, the vessels are provided for the facilities to exclude the direct gas-jet effect on the fluid surface and to provide a uniform gas supply to the vapor space of the vessels. Moreover, special measures are taken to exclude formation of a hopper above the outlet connections.

In spherical and vertical cylindrical vessels, for supercharging gas input, use is made of a collector in the form of an annular perforated tube, its perforation holes being located on the upper generatrix (Fig. 14.6a). This provides gas efflux in the form of separate jets vertically upward and their deceleration near the dome walls. A similar scheme of collector for vapor-space supercharging but in the form of a horizontal tube with holes on the upper generatrix and with dead ends is used in horizontal cylindrical vessels (Fig. 14.6b).

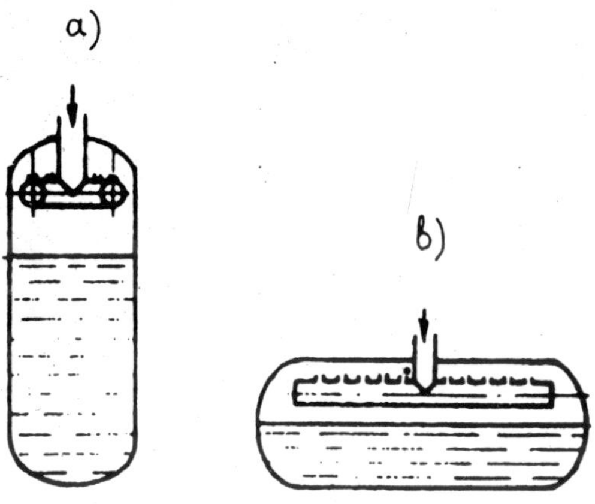

Fig. 14.6. Typical gas supply collector designs: a, in vertical, cylindrical and spherical vessels; b, in horizontal cylindrical vessels.

Cryogenic pipelines for transporting liquid and gaseous product must maintain safe operation characteristics over the entire operating temperature range from +50°C to temperatures of products to be transported, and also give a sufficiently small environmental heat flux. Meeting these requirements is attained by: choosing materials capable of working at low temperatures; making decisions on compensating stresses that appear on pipeline cooling; effective thermal protection from environmental heat fluxes.

Modern commercial designs of cryogenic pipelines are based on vacuum-type insulations: vacuum-powder, layered-vacuum, and sometimes pure vacuum. Designs of smooth seamless or welded pipelines have found most use in cryogenic systems. Stainless steel and sometimes Invar are

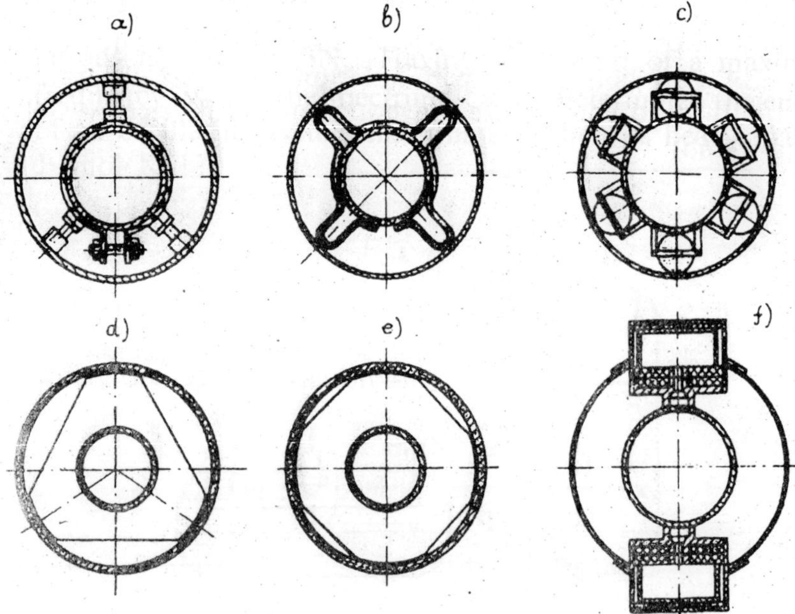

Fig. 14.7. Designs of inner tube supports relative to the shell: a, finger; b, wire; c, ball; d, e, disc; f, pivot.

used as structural materials of an inner tube. An outer shell is manufactured from stainless or carbon steel. When Invar is used as the structural material of the inner tube the problem on compensating temperature stresses under pipeline cooling is solved more simply, since the mean linear expansion coefficient of Invar is six times less than that of stainless steel.

The space between the inner tube and shell is evacuated, and so that the thermal protection efficiency is increased it is filled either with a powder (aerogel or perlite) or else used with vacuum-layer insulation in it. The inner tube is fixed relative to the shell by supports: finger, wire, ball, disc, and pivot (Fig. 14.7). Wire, finger and ball supports are intended only to provide co-axial alignment of the inner tube and shell, while the disc and pivot supports also provide transfer of inner-tube forces to the shell. A length of pipeline section with one vacuum cavity has the advantages of easy vacuum treatment, and simple leaks tracing, transportation and mounting. Vacuum cavities of separate pipeline sections are closed by thermal bridges that provide a hardwired connection of the inner tube and shell. Several-type bridge designs have been developed: telescope, cylinder, cone, and bellows (Fig. 14.8). Because of the high cost, the latter are not very often used. Telescope, cylinder and cone bridges have found most use: the first, because of their good thermal characteristics, the second because of their simple design. In a number of cases, thermal bridges are intended only to provide a hardwired connection between inner and outer tubes, and in this case these are manufactured perforated.

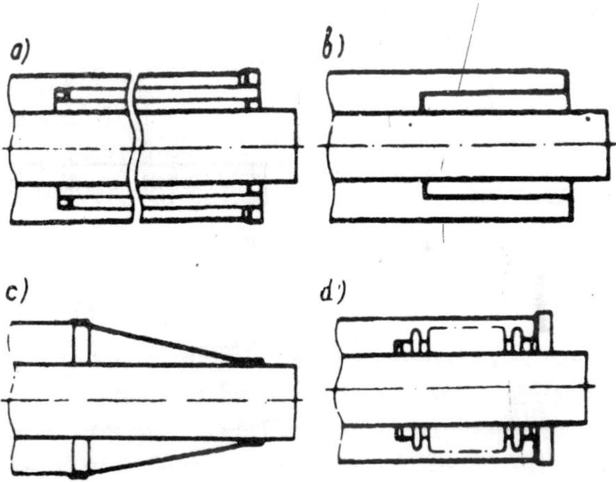

Fig. 14.8. Design schemes of thermal bridges; a, telescope-type; b, cylindrical; c, cone; d, bellows.

Joining separate pipeline sections with vacuum cavities is achieved by welding or flange connections. The main design feature of flange connections of cryogenic pipelines is that a unit mounting the sealing gasket is placed into the heated zone. Through this, reliable sealing of the connections after inner tube cooling can be provided. Among the known designs, the so-called bayonet joints (Fig. 14.9a) have found most use and, in addition to their ability to offer good sealing, these also provide sufficiently small heat fluxes to a product to be transported.

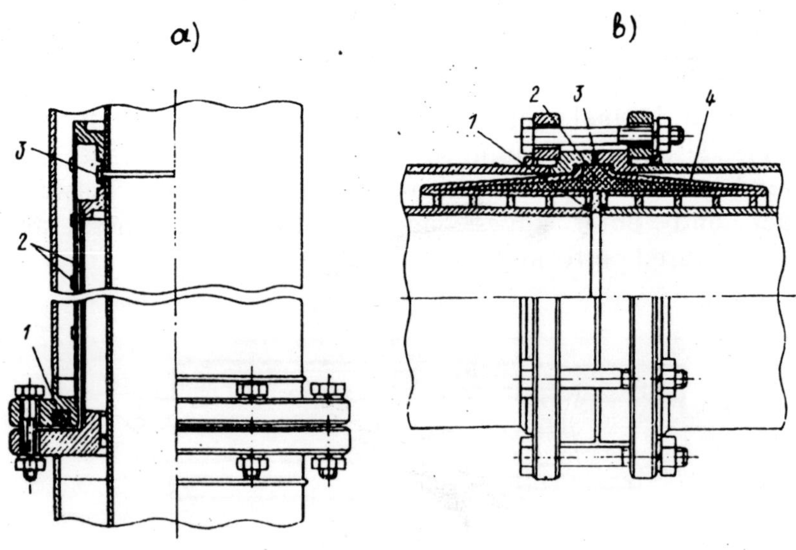

Fig. 14.9. Designs of flange detachable connections of cryogenic pipeline sections with independent insulation cavities: a, bayonet joint (1, rubber ring; 2, thermal bridge; 3, safety ring); b, single-plane connection (1, safety ring; 2, insulation; 3, outer seal; 4, thermal bridge).

A single-plane connection (Fig. 14.9b) possesses slightly worse thermal characteristics but enables the dismounting of separate sections to be simplified. Flange connections are used in cryogenic pipelines where frequent re-mounting of equipment is needed. In the remaining cases, more reliable welded designs (Fig. 14.10) are used.

Expansion joints are provided for in a pipeline design to eliminate temperature stresses that appear under inner-tube cooling; in the case of considerable ambient temperature fluctuations, the expansion joints are also mounted on the outer shell. As stress expansion joints use is made of different elastic elements: bellows, corrugated hoses, lens compensators. Sometimes the elastic properties of the elbows of the pipelines themselves are used to compensate temperature stresses.

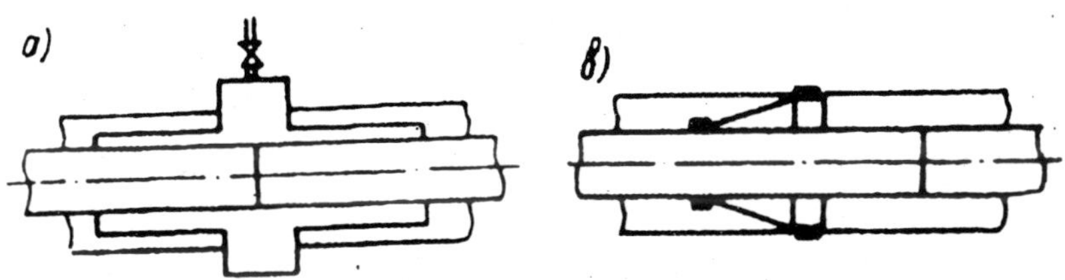

Fig. 14.10. Designs of welded joins of sections of cryogenic pipelines with independent insulation cavities: a, independent transportable sections; b, sections to be evacuated after mounting.

It is advisable to mount liquid cryoproduct pipelines with a small lift in the fluid direction. Subsequently, the vapor hampers the fluid flow less, and after the valve is closed or the outlet pipeline end, the vapor displaces the remaining fluid back to the vessel.

Liquid cryoproduct transfer via the pipelines is associated with the possible appearance two-phase flow due to evaporation of some amount of fluid caused by the environmental heat flux. This results in substantially decreased pipeline capacity.

The sound velocity that characterizes the flow velocity limit in single-phase low-temperature fluids is rather high (approx. 1000 $m \cdot s^{-1}$ for nitrogen, oxygen and hydrogen and approx. 200 $m \cdot s^{-1}$ for liquid helium) but is very much decreased in a two-phase mixture owing to high vapor compressibility. The process is complicated by mass exchange between the vapor and fluid, and non-equilibrium state phase. It is therefore advisable to create conditions that exclude two-phase flow onset (system pressure elevation, fluid cooling, heat flux decrease, etc.).

Initial cooling of a cryopipeline is a separate and difficult problem. To reduce cooling time of long pipelines, the gas discharge valves or special gas fluid separators are mounted at certain intervals, thereby avoiding the pumping of the gas troughout the entire pipeline, and allowing fluid front propagation to advance.

Questions

1. What determines the shape of a cryogenic vessel?
2. What are the methods of decreasing heat fluxes to a cryogenic vessel?
3. How is the thermal protection efficiency of a cryogenic vessel calculated?
4. What are the designing features of helium vessels?
5. What are the main features of cryoproduct storage under space-flight conditions?
6. What difficulties appear in storing liquid fluorine?
7. What are the main features of cryogenic pipelines?

15

PROCESSES IN CRYOGENIC VESSELS

15.1. Features of processes in cryogenic vessels

Certain aspects of storing cryogenic products in vessels are connected with variation in their parameters as a result of environmental heat flux. The available heat flux results both in product evaporation and loss if the vapor space of vessels is connected to the environment or in an increase in internal energy and hence pressure and temperature, for products stored in closed vessels. The main advantages of liquid cryoproducts as against gaseous ones (larger density and large supply of cold in the form of latent heat of vaporization) are most fully realized during storage at pressures close to atmosperic, i.e. with discharge of vapors forming.

The time of vapor discharge storage in modern vessels is sufficiently long; relative fluid losses through evaporation are usually less than 1% a day. At the same time, such a small evaporation rate in many cases makes it inapplicable for liquid cryogenic product open-vessel storage with gas discharge, because a product may be contaminated and air may leak into the inner cavities.

Under storage conditions, the amount of vapors forming per unit time is so small that the vessel pressure practically corresponds to atmospheric. As a result, with decrease of atmospheric pressure, some amount of the product boils up, and with elevating the pressure the air flows via a drainage communication into the vessel volume, with air components partially condensing on the fluid surface.

In practice, realization of a particular phenomenon depends on the rise-rate ratio of an atmospheric pressure $\mathbf{dp_a/d\tau}$ and rise-rate ratio of vessel pressure $\mathbf{dp/d\tau}$ for closed-vessel storage with gas discharge; if

$$\frac{dp_a}{d\tau} > \frac{dp}{d\tau}$$

then air leak-in occurs.

To estimate the pressure rise-rate in a closed vessel, use is made of the equation for pressure change on the saturation line, i.e. the Clapeyron-Clausius equation and the heat balance equation for mass of product to be stored; in this case, the last equation neglects the vessel vapor mass, which is quite admissible for high of degrees filling:

$$\left(\frac{dp}{dT}\right)_s = \frac{r}{T(v'' - v')}; \tag{15.1}$$

$$Q\, d\tau = M\, c_s\, dT\,. \tag{15.2}$$

Here **M** is the fluid mass; c_s is the specific heat capacity of fluid on the saturation line; **Q** is the heat flux (watts) supplied to the fluid; **r** is the specific evaporation heat; **v''** is the specific volume of vapor; **v'** is the specific volume of liquid.

The heat flux **Q** may be expressed in terms of relative fluid losses per unit time η

$$\eta = \frac{-1}{M}\frac{dM}{d\tau} = \frac{Q}{rM}\,. \tag{15.3}$$

Simultaneously solving eqns. (15.1) ÷ (15.3) gives a formula for estimating a vessel pressure rise-rate for closed-vessel storage with gas discharge

$$\frac{dp}{d\tau} = -\frac{\eta r^2}{c_s T(v'' - v')}\,. \tag{15.4}$$

Analysis of this expression shows that with improved quality of thermal protection of vessels (decreasing η), the probability of air leak-in grows. So, with a daily 0.1% evaporation of liquid oxygen, air leak-in will already occur with an atmospheric pressure rise-rate above 50 $Pa \cdot h^{-1}$. Such an atmospheric pressure rise-rate may occur when varying both atmosphere state and ground altitude during transportation of a product vessel.

To avoid air leak-in, in a number of systems the cryogenic product storage with vapor discharge is arranged with some excess pressure. This is achieved by mounting hydraulic back-pressure valves, special drainage valves or gas discharge small-capacity lines. Sizes and hydraulic characteristics of such lines are calculated so that the excess pressure Δp would exceed a fluctuation in amplitude of atmospheric pressure Δp_a .

It should be noted that storage under small excess pressure with continuous gas discharge complicates the design and service of storage and is used in very large low-pressure vessels, e.g. for storage of natural gas and air separation products.

Storage with periodic gas discharge over some excess pressure range has found wide use. The upper pressure limit corresponding to the start of gas discharge is confined to the vessel pressure while the lower one corresponds to the gas discharge end, facilitates pressure control, and usually amounts to several hundred fractions of a MPa. Such storage technology is convenient to use and requires no additional modifications of vessel design since all vacuum insulation-equipped vessels have a working pressure not less than 0.17 MPa, and modern thermal protection allows sufficiently long storage without vapor discharge. Closed-vessel storage whith gas discharge enables a working product to be saved if it is used periodically in the form of fluid or gas.

In addition to the advantages of providing product purity, storage with blind drainage also has some disadvantages that can complicate the operation of a system. For closed-vessel storage with gas discharge, the useful vessel volume somewhat decreases as the vapor space above the fluid surface must be sufficient to compensate fluid thermal expansion; moreover, heating of a vessel fluid causes it to steam in lines on gas discharge, thereby reducing product flow-rate and upsetting pump operation. A limited storage time must be taken into account during long-distance transportation of cryogenic products.

15.2. Calculation of cryoproduct parameters change with storage

Calculating the parameters of cryoproducts to be stored in closed vessels with gas discharge presents no difficulties if the product temperature in the vessel is uniform and varies only in time.

It is known that with heat supplied, in the isochoric process the internal energy of a thermodynamic system obeys the relation

$$\mathbf{M}\,(\mathbf{u} - \mathbf{u}_0) = \mathbf{Q}\,\tau\,, \qquad (15.5)$$

where **M** is the product mass in a vessel; **Q** is the environmental heat flux; $\mathbf{u}_0$, $\mathbf{u}$ are the specific internal energies of the product at start and end of storage, respectively.

The specific internal energy of a two-phase system may be expressed in terms of a specific internal energy of each of the phases:

$$\mathbf{u} = \frac{\mathbf{u'M'} + \mathbf{u''M''}}{\mathbf{M'} + \mathbf{M''}} = \mathbf{u'}(1 - \varphi)\,\frac{\mathbf{v}_m}{\mathbf{v'}} + \mathbf{u''}\varphi\,\frac{\mathbf{v}_m}{\mathbf{v''}}\,, \qquad (15.6)$$

where **M', M''** are the mass of fluid and vapor in a vessel; **v', v''** are the specific volume of fluid and vapor phases; **u', u''** are the specific internal energy of the fluid and vapor phase, respectively; $\mathbf{v}_m$ is the specific volume of a product; φ is the vapor volume content.

The specific volume of a product in a vessel is calculated in terms of initial specific volumes of each of the phases

$$v_m = \frac{V}{M' + M''} = \frac{v_0' v_0''}{v_0''(1 - \varphi_0) + v_0' \varphi_0}, \tag{15.7}$$

where **V** is the vessel volume; v_0', v_0'' are the specific volumes of the phases at the storage start; φ_0 is the volume fraction of the vapor at the storage start.

Since in the isochoric process the specific volume of a product remains invariable, expression (15.7) implies that a current value of a volume vapor content is determined by the relation

$$\varphi = \frac{v'' v_m - v' v''}{v'' v_m - v' v_m}. \tag{15.8}$$

Taking account of relation (15.8), the expression for specific internal energy of a product, (15.6), assumes the form

$$u = \frac{u'(v'' - v_m) + u''(v_m - v')}{v'' - v'}. \tag{15.9}$$

Fig. 15.1, Fig. 15.2, and Fig. 15.3 show plots of calculations of pressure variation for liquid oxygen, hydrogen, and helium versus heat flux to unit volume of a product for different initial volume vapor contents. Breakpoints correspond to complete filling of the vessel with fluid due to its temperature expansion.

Fig. 15.4 shows design curves for relative fluid volume **(1 – φ)** variation nitrogen no-drainage storage. Varying the relative volume of fluid to be stored takes place differently for different values of initial filling degree **(1 - φ₀)**. If the initial volume vapor content φ_0 is sufficiently small, then because of thermal expansion, as heat is supplied, the fluid occupies more of the vessel volume; in this case, the vapor

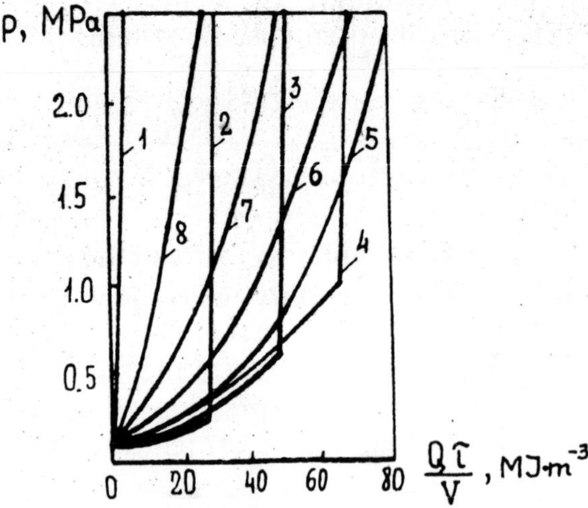

Fig. 15.1. Pressure increase under thermodynamic equilibrium storage of liquid oxygen vs heat supplied to unit volume of the vessel for different initial volume vapor fractions: 1, φ_0= 0; 2, 0.05; 3, 0.1; 4, 0.15; 5, 0.3; 6, 0.5; 7, 0.7; 8, 0.9.

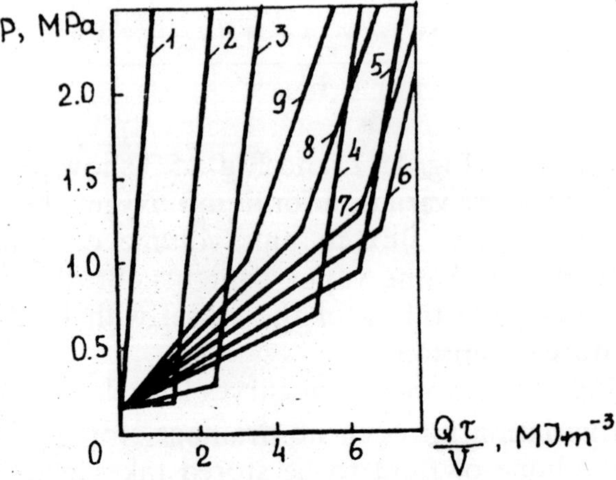

Fig. 15.2. Pressure increase under thermal equilibrium no-drainage storage of liquid hydrogen vs heat supplied to unit volume of the vessel for different initial volume vapor fractions: 1, φ_0= 0; 2, 0.05; 3, 0.084; 4, 0.2; 5, 0.285; 6, 0.4; 7, 0.58; 8, 0.7; 9, 0.81.

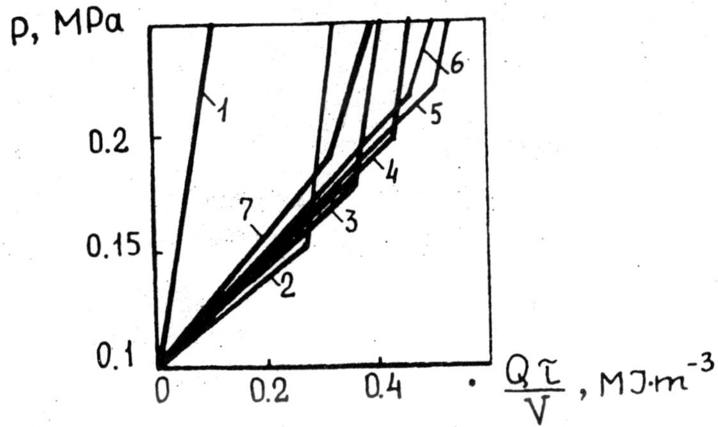

Fig. 15.3. Pressure increase under thermodynamic equilibrium liquid helium storage vs supplied heat based on unit volume of the vessel for different initial volume vapor fractions:
1, φ_0= 0; 2, 0.106; 3, 0.154; 4, 0.215; 5, 0.32; 6, 0.54; 7, 0.686.

condenses on the fluid surface. For large initial volumes of vapor-phase fluid, thermal expansion is compensated by its evaporation, with heat supplied. As a result, the volume gradually reduces,and the fluid may completely evaporate. Analysis of the pressure-rise curves in Fig. 15.1, Fig. 15.2, and Fig. 15.3 shows that with increase of pressure, the initial volume fraction of fluid to be stored up to an assigned pressure first increases and then slows down. This characteristic pattern is caused by the fact that, though increasing a fluid fraction in a vessel gives an increase in system heat capacity and a pressure rise-rate decrease in a two-phase region, after having filled the vessel with fluid, the pressure-rise rate increases sharply owing to fluid thermal expansion; in this case, the higher the filling degree of a vessel, the earlier the volume is filled with liquid phase.

Fig. 15.5 plots, as an example, experimental data and time variations of liquid nitrogen no-drainage storage up to

0.5 and 0.6 MPa in two industrial vessels. Comparison of the predicted and experimental values implies that the latter qualitatively support the predicted results; however, a quantitative disagreement is attributed to the upper fluid layer temperature growing more rapidly than that of deep layers.

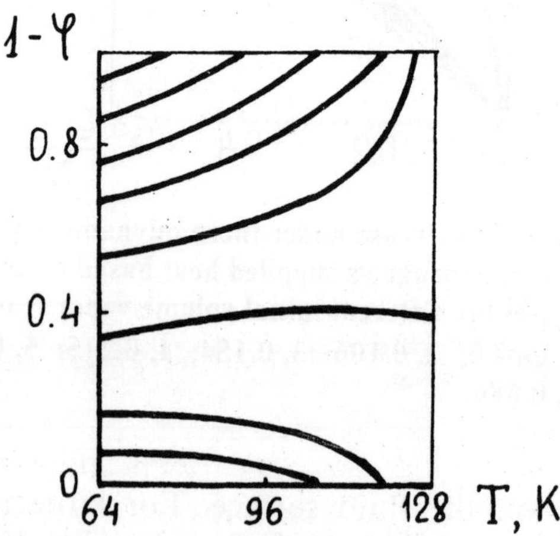

Fig. 15.4. Temperature variation of relative liquid nitrogen volume (1 - φ) under closed-vessel storage.

Schlieren studies of thermal phenomena to be visualized due to optical non-uniformity of a medium have led to the discovery of a physical pattern of a temperature stratifica-tion onset. It is found that with side heat supply, a free convective layer is formed near the vessel walls and its thickness increases over its height. A considerable amount of side heat flux is accumulated by this layer and is carried away in the form of heated fluid to the phase interface. Also, the upper heat flux favors a temperature stratification since, causing the upper fluid layer heating, it does not provide convective mixing conditions. In real vessel designs, the temperature field development in cryogenic products is

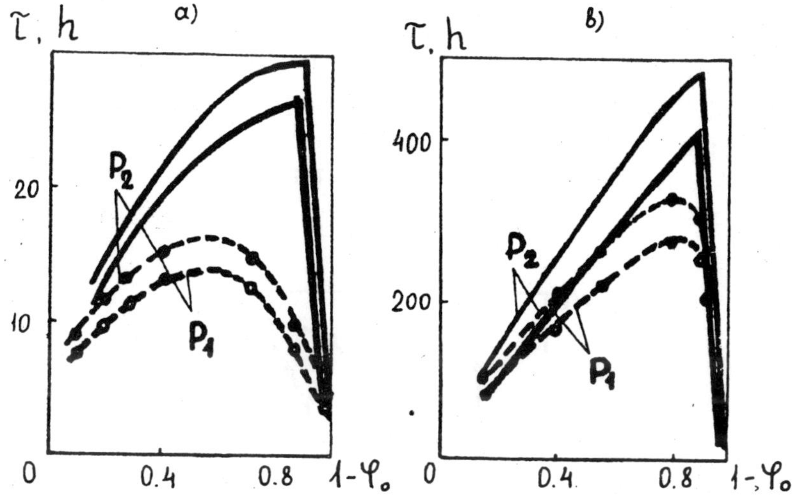

Fig. 15.5. Vessel pressure increase time under liquid nitrogen storage with closed gas discharge from p_0= 0.1 MPa to p vs filling degree (p_1=0.5 MPa, p_2=0.6 MPa): a, vessel with V = 14 m^3; b, vessel with V = 2.6 m^3; solid lines, calculation assuming thermodynamic equilibrium; ○, ● - experiment.

affected by non-uniform heat supply over the vessel surface, heat flux value, vessel filling degree, phase transitions on the fluid surface, thermophysical properties of a product. Typical experimental curves for liquid nitrogen temperature variations over vessel height at different time instants are plotted in Fig. 15.6.

The temperature curve shows that a zone with a practically uniform temperature (the so-called core) is formed in the bottom part of the fluid volume; the upper stratified fluid layer is heated more than the core, the fluid temperature increasing as it approaches the interface. The temperature difference on the fluid surface and in the core ΔT is called the value of temperature stratification.

Fig. 15.7 plots experimental temperature fields over

liquid nitrogen depth for no-drainage storage, illustrating the heat flux effect per unit time, all other conditions being equal. These temperature curves imply that on increasing the density of the heat flux to a product, the stratified layer thickness decreases and the temperature stratification increase.

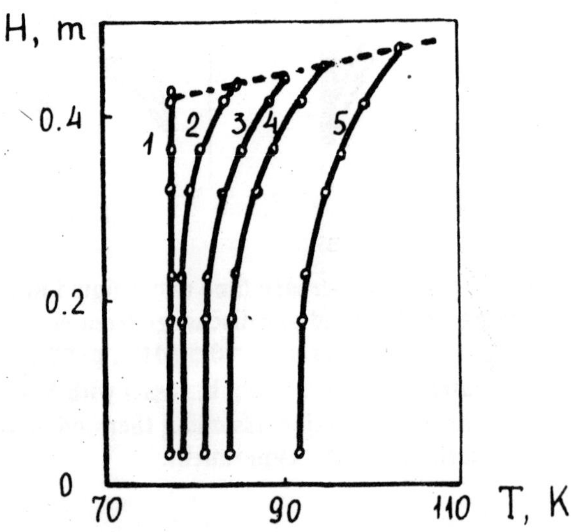

Fig. 15.6. Experimental curves for variation in height of liquid nitrogen temperature at different time instants, starting with no-drainage storage in the vessel with V = 0.1 m^3, (1 - φ) = 0.77: 1, τ = 0; 2, 12 h; 3, 26 h; 4, 46 h; 5, 84 h.

Fig. 15.8 plots experimentally time-measured temperature stratification for different values of initial fluid fraction. The temperature stratification grows with time, its growth rate gradually decreasing. For low degrees of filling, the temperature stratification is inconsiderable because of convective mixing affected by bottom heat flux; moreover,

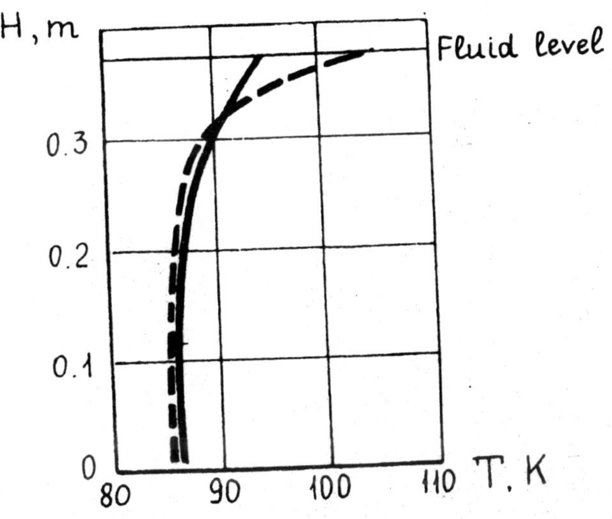

Fig. 15.7. Experimental height profiles of liquid nitrogen temperatures with the same total amount of heat supplied $Q\tau$ = 1140 kJ in the vessel with V = 0.1 m^3. Filled fluid mass M = 50 kg, filling degree (1 - φ_0) = 0.62, initial temperature T_0= 77 K.
——— - environmental heat flux Q = 8.1 W;
- - - - environmental heat flux Q = 380 W.

fluid evaporation from the surface favors temperature equalization.

The temperature stratification of cryogenic products not only hinders parameter calculation but also deteriorates vessel performance characterictics, since it causes more rapid pressure growth compared with uniform heating (Fig. 15.9).

Methods to avoid temperature stratification are based on mixing of fluid to be stored. To mix fluid, some devices are proposed. The main designs are shown in Fig. 15.10. In jet mixers (Fig. 5.10a, b), mixing is achieved by ascending fluid jets formed either by mechanical means or by special

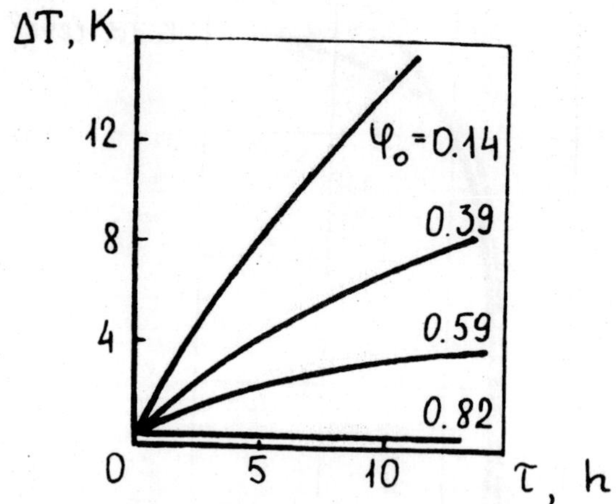

Fig. 15.8. Surface and fluid core temperature difference vs storage time for different initial filling. Working fluid, nitrogen; T_0= 77 K; horizontal vessel with V = 2.6 m^3; D = 1.2 m; Q = 32 W.

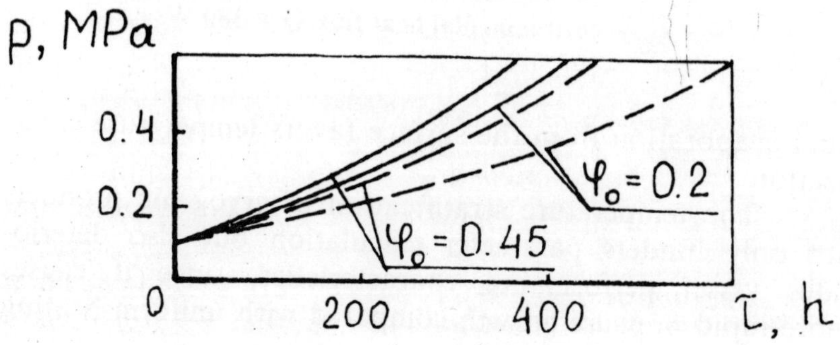

Fig. 15.9. Pressure change under no-draining storage of liquid nitrogen in a horizontal cylindrical vessel with V = 2.6 m^3 and Q = 28 W: dotted lines, calculation for equilibrium storage; solid lines, experiment.

heaters. Fig. 15.10c shows a design with a temperature-stratification reduction due to a redistribution of heat flux to the vessel bottom part. Redistribution is achieved by placing, in the insulation space, a heat-conduction thermal-contact screen in the bottom part of the inner vessel. Device operation (Fig. 15.10d) is based on vapor-phase condensation in a special vessel placed into the lower fluid layers, thereby reducing the temperature stratification.

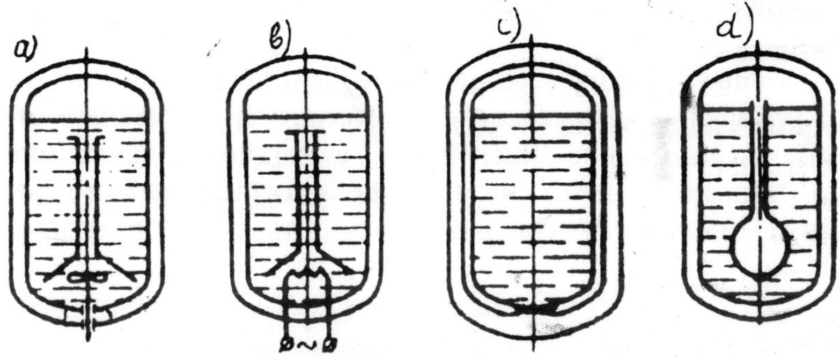

Fig. 15.10. Cryoproduct device for decreasing the temperature spread: a, mechanical mixer; b, thermal mixer; c, insulation cavity screen closed to the vessel bottom for redistributing the heat flux; d, small tank for heat removal from the fluid surface through to vapor condensation.

Calculating cryogenic product parameters with no-drainage storage requires solving a system of differential heat transfer and hydrodynamics equations describing natural convection in a closed volume. In the last 15 ÷ 20 years, through the wide use of fast-response computers, numerical solutions of the problem are obtained in the form of temperature and velocity fields in fully or partially fluid-filled volumes. However, numerical methods of solving the main system of equations are possible only for laminar fluid flow corresponding to the Rayleigh number $\mathbf{Ra} < 10^9$, where

$$\mathbf{Ra} = \frac{\mathbf{q}\,\mathbf{H}^3}{\mathbf{a}_l\,\nu_l}\beta\,\Delta\mathbf{T}$$

In commercial vessels, turbulent fluid flow (**Ra** » 10^9) occurs, thus preventing the use of results obtained in practice. Consequently, a quantitative description of no-drainage storage uses approximate methods based on simplified physical models or experimental data processing.

The boundary layer model is most consistent with the real physical nature of form the formation of temperature fields. According to this model, a boundary layer that transports hot fluid to the interface is formed near the side walls of the vessel. A dome heat flux also favors upper layer heating.

Fig. 15.11 shows fluid flow pattern and temperature fields which clarify the physical model and illustrate the meaning of some equations when applied to a vertical cylinder.

For relatively high heat flux densities near the vessel surface (**q** » 1 W·m^{-2}), with vapor space volume being equal to $\varphi = 0.1 \div 0.2$, the design estimates of a fluid surface tempe-rature may use a simple physical model, according to which a temperature spread arises owing to heat supply to the fluid surface. It is assumed that the heat flux to a fluid surface enters from the top of the vessel; the heat flux from the fluid-wetted walls is used in unifirm fluid heating. The fluid temperature is calculated by summing temperature fields due to heat fluxes through an interface and through vessel walls contacting the liquid product.

The heat flux supplied across fluid surface is assumed to spread in the liquid as in a semi-infinite body with effective thermal conductivity

$$\lambda_{el} = \varepsilon\,\lambda_l$$

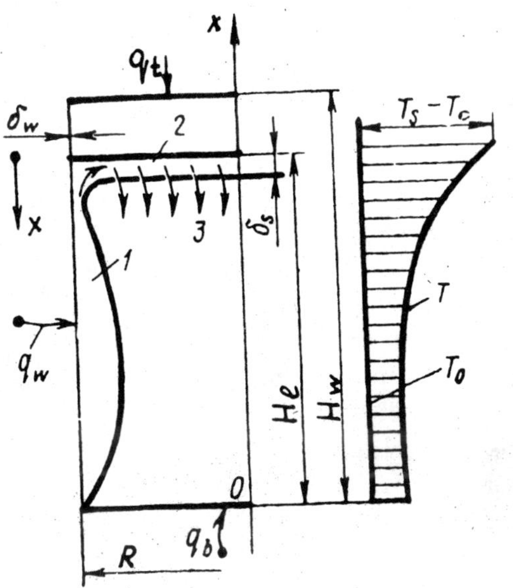

Fig. 15.11. Schematic representation of typical flow zones and fluid temperature distribution according to the boundary layer model: 1, boundary layer; 2, surface layer; 3, core.

where ε is the coefficient taking into consideration the liquid convection:

$$\varepsilon = \mathbf{B}\ \mathbf{Ra}^{1/6}\ .$$

Here **B** is the constant: **B** = 0.04 for a sphere and **B** = 0.08 for a vertical cylinder; **Ra** is modified Rayleigh number:

$$\mathbf{Ra} = \frac{g H_l^4 \beta q}{a_l \nu_l \lambda_l}\ .$$

In such a problem, the temperature field in the vessel is determined by the expression:

$$T(x,\tau) = T_0 + \frac{q F_l \tau}{\rho_l c_{pl} V_l} + \frac{q F_v}{F_s} f(x,\tau) \ , \qquad (15.10)$$

where

$$f(x,\tau) = 2\sqrt{\frac{\tau}{\rho_l c_{pl} \lambda_{el}}} \ \text{ierfc} \frac{x}{\sqrt{a_{el} \tau}} \ ;$$

T_0 is the initial temperature; **q** is the heat flux density (**q** is assumed constant along all the vessel surface); F_l is the wall surface moistened with liquid; F_v is the wall surface adjacent to vapor; F_s is the liquid-vapor interface; V_l is the liquid volume; **x** is the vertical coordinate.

From eqn. (15.10) it follows that the fluid surface temperature is calculated by the formula

$$T_s = T_0 + \frac{q F_l \tau}{\rho_l c_{pl} V_l} + 1.13 \frac{q F_v}{F_s} \sqrt{\frac{\tau}{\rho_l c_{pl} \lambda_{el}}} \ . \qquad (15.11)$$

Engineering estimates of pressure rise often use the empirical relations obtained from experimental data processing. For vertical cylindrical vessels, the relation

$$p - p_0 = K_0 q \tau \frac{D^{1.03} \lambda_w^{0.3}}{F^{1.46} \delta_w^{0.72}} (H_l / D)^{-n} \qquad (15.12)$$

may be used where K_0 is equal to 5.2×10^{-4} for O_2, 6.5×10^{-4} for N_2, 1.85×10^{-4} for H_2, 2.86×10^{-2} for He. A value of the power **n** in formula (15.12) is determined depending on a liquid layer height H_l / D (Fig. 15.12). Dimensions of values in formula (15.12) are: **p**, p_0 - atm; τ - h; **D**, δ - m; **F** - m^2; **q** - kcal·(m^2·h)$^{-1}$; λ_w - kcal·(m·h·grad)$^{-1}$.

Ranges of quantities, to which relation (15.12) is extended, are:

$$(p - p_0)(v'' - v')/r = 10^{-3} \div 10;$$

$$q D \beta_l / \lambda_l = 2.5 \times 10^{-2} \div 8.5;$$

$$v''/v' = 8 \div 250; \quad \lambda_w / \lambda' = 3.5 \div 64; \quad H_l / D = 0.5 \div 2.2;$$

$$\delta_w / D = 2\times10^{-3} \div 1.5\times10^{-2}; \quad a_l \tau / D^2 = 10^{-3} \div 0.1.$$

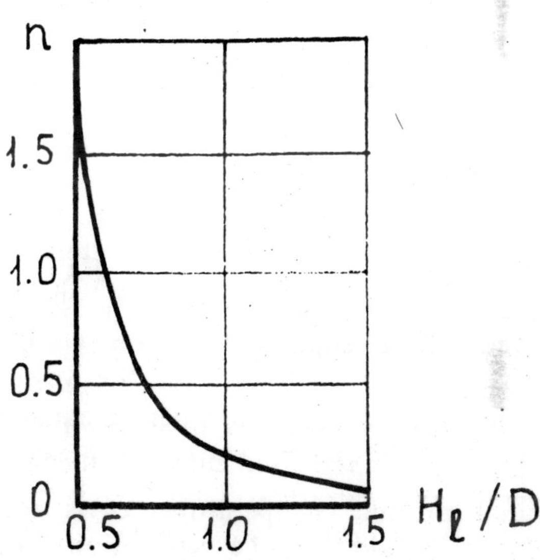

Fig. 15.12. Value of the power n vs fluid layer heigh H_l/D in formula (15.12)

Also, the proposed relation may be adopted for estimating the pressure rise in horizontal cylindrical vessels for high degress of filling $H_l / D_{ef} > 0.45$; in this case, the hydraulic diameter calculated for a longitudinal axial cross-section may be taken as a determining dimension; the power **n** at factor $\mathbf{H_l / D_{ef}}$ in eqn. (15.12) should be taken equal to 1.

To calculate the time of no-drainage storage of liquid nitrogen, argon, and oxygen in commercial vessels, preference is given to the empirical relations checked over a wide range of operating parameters.

Calculations use the following dimensionless variables

$$\upsilon = \tau_p / \tau\,; \quad \psi = (p - p_0) / (p' - p_0); \quad \varphi_0 = V_0'' / V;$$

$$\zeta = \frac{q F \tau^*}{M_0 c_0 T_0},$$

where τ is the time from the beginning of closed-vessel storage with gas discharge; τ_p is the design storage time up to an assigned pressure under thermodynamic equilibrium; **p** is the assigned ultimate pressure; p_0 is the initial pressure; **p'** is the design pressure at which a two-phase system changes to a single-phase state; φ_0 is the vessels initial volume vapor content; V_0'' is the initial [illegible] of the vapor space; **V** is the vessel volume; [illegible]s the heat flux density based on the vessel surface; [illegible] the vessel surface; τ^* is the time period equal to 1 hour; $M_0 = \rho_l V (1 - \varphi_0)$ is the fluid mass at start of storage; T_0 is he initial product temperature; c_0 is the specific heat capacity of liquid on the saturation line at the temperature T_0.

Design formulas are of the form:

- for pressure range $\psi = 0.05 \div 1.0$

$$\upsilon = [1.59 + 0.83\,(0.5 - \psi) + 0.83\,(0.5 - \psi)^2\,] \times$$

$$\times\,[0.8 + 0.236\,(1 - \varphi_0 - 0.35) + 0.32\,(1 - \varphi_0 - 0.35)^2 +$$

$$+ 4.38 (1 - \varphi_0 - 0.35)^3] \times$$

$$\times [0.8 + 20 (\varsigma - 0.001) - 204 (\varsigma - 0.001)^2] ;$$

- for pressure range $\psi = 1.0 \div 4.0$

$$\upsilon = [0.64 + 0.046 (3.5 - \psi) + 0.031 (3.5 - \psi)^2 +$$

$$+ 0.026 (3.5 - \psi)^3] \times [0.8 + 0.236 (1 - \varphi_0 - 0.35) +$$

$$+ 0.32 (1 - \varphi_0 - 0.35)^2 + 4.38 (1 - \varphi_0 - 0.35)^3] \times$$

$$\times [0.8 + 20 (\varsigma - 0.001) - 204 (\varsigma - 0.001)^2] .$$

These relations are extended to cylindrical and spherical vessels where relative liquid-phase volume amounts to $(1 - \varphi_0) = 0.2 \div 0.9$; heat flux density is $\mathbf{q} = 5 \div 20$ W·m^{-2}; initial pressure is equal to atmospheric $p_0 = 0.1$ MPa.

Questions

1. Under what conditions does air leak into the cryo-fluid vessel? How can this phenomenon be avoided?
2. How is the rate of vessel pressure increase for closed gas drainage calculated?
3. Derive the formula for a current value of the volume vapor content for fluid storage with closed gas drainage.
4. How does the temperature stratification of a cryo-fluid affect vessel pressure increase?
5. By what formula is the surface fluid temperature calculated?

16

CRYOGENIC SYSTEMS FOR SCIENTIFIC RESEARCH

Use of low and ultralow temperatures opens up exclusive possibilities for fundamental researchs. Cryogenic devices and plants are important for supporting experiments in the fields of elementary-particle physics, nuclear power engineering, solid-state physics, etc.

16.1. Cryostats

In the laboratory, cryostat-flasks for objects to be thermostated at cryotemperatures are widely used. Usual cryoproducts are liquid helium, hydrogen, nitrogen, etc., and, in some cases, also solid cryoagents to be poured into a cryostat serve as a cold source.

By varying the vapor pressure above the fluid poured into the cryostat, the fluid temperature may be altered from the critical point to the triple point. In practice, it is convenient to have a pressure above the fluid approximately equal to atmospheric or below (by pumping out its vapors). As a result, temperatures may be obtained and easily maintained over the following ranges (Table 16.1).

Table 16.1. Temperature ranges of cryoproducts.

Temperature range, K	Cryoproduct
90 ÷ 55	Oxygen
78 ÷ 63	Nitrogen
27 ÷ 24.5	Neon
20.4 ÷ 14	Hydrogen
4.2 ÷ 1.0	Helium - 4
1 ÷ 0.3	Helium - 3

Cryotemperatures between the above ranges and, also above 90 K, are achieved by heating a cold gas of the corresponding cryoproduct or by a heater-equipped coolant line.

Temperatures below the triple point may be attained by cryoproduct sublimation, i.e. direct transition from solid to vapor state.

Sublimation cooling systems are primarily of promise for space investigations. These sublimation cryostats use as working substance solid hydrogen, neon, nitrogen, argon, methane, carbon oxide, ethylene, solid carbon dioxide and ammonia, to achieve temperatures above 120 K.

Temperatures to be attained with solid cryoproduct sublimation (from the triple point to a temperature corresponding to vapor pressure over the crystal approximately equal to 20 ÷ 100 Pa) lie within the following ranges: for hydrogen, 14 to 8 K; for neon, 25 to 14 K; for nitrogen, 63 to 42 K; for argon, 84 to 48 K; for methane, 91 to 60 K; for ammonia, 195 to 148 K; for carbon dioxide, 216 to 125 K.

When designing any type of cryostat, the primary task is to protect the cryostat volume from environmental heat fluxes. The lower the boiling point and evaporation heat of a working cryoagent the more difficult the task.

For cryostats to be poured with liquid nitrogen or oxygen, good results are given by high-vacuum thermal insulation similar to that used in the common Dewar flask or domestic thermos.

For helium cryostats, ordinary high-vacuum thermal insulation does not yield satisfactoty results since the heat radiation of outer walls remains substantial compared to small specific evaporation heat of liquid helium.

Radiant heat fluxes for a particular surface emissivity may be lowered by the following methods:

(1) reducing the temperature of emitting walls (cooling the walls by an auxiliary cryoagent or evaporating cryoproduct);

(2) mounting shielding screens (cooled or not);

(3) utilizing multilayer screen-vacuum thermal insulation.

Fig. 16.1 shows evaporation rates of liquid helium from silvered glass Dewar flasks for three cases: a, outer wall and cover have normal temperature (≈ 300 K); b, outer wall cooled by liquid nitrogen; c, outer wall cooled by liquid nitrogen and cover emission screened by a 5-cm-thick cellular plastic plug suspended at a distance of 12 cm from the top flask edge.

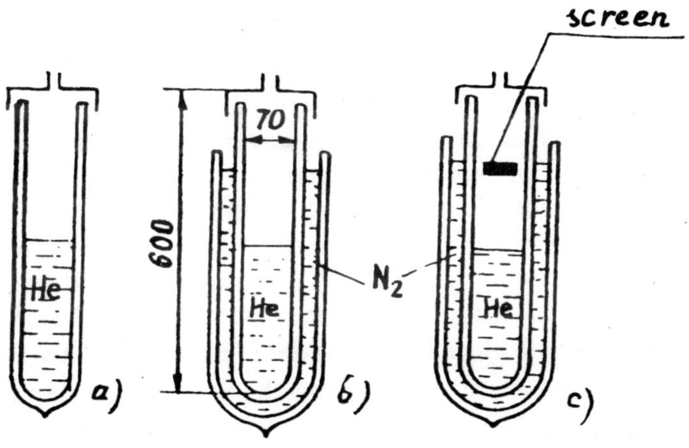

Fig. 16.1. Evaporation rates ($cm^3 \cdot h^{-1}$) of liquid helium from glass Dewar flasks: a, 2500; b, 60; c, 45.

As well as radiation, heat supply from the environment may occur for the following reasons: heat conduction of gas left in the vacuum insulation space (or heat conduction of material filling this space); heat conduction of pipes, supports, electric wires and other joints between parts of a cryostat at a different temperature.

Heat supply to the fluid may also be produced by the Joule heat or heat to be released due to the appearance of Foucault currents; mechanical oscillations if the latter originate in the cryostat itself or are transferred to it from the outside.

The main method of reducing liquid helium losses by different heat bridges is liquid nitrogen and evaporating helium cooling of these bridges.

A substantial decrease in evaporation rate of liquid helium from the environmental heat fluxes may be achieved by cryogenic gas machines used for cooling screens, supports, etc.

16.1.1. Glass cryostats

Glass cryostats are widely used under laboratory conditions because they are simple in manufacture and allow an experiment's progress to be seen.

The schematic of a typical helium cryostat is shown in Fig. 16.2. The cryostat is composed of two Dewar flasks, one inserted into the other. Liquid helium is poured into the inner flask and liquid nitrogen into the outer one. The surfaces of both flasks are silvered, except for two vertical $10 \div 15$-mm-wide strips along the entire flask length for allowing direct observation. The helium flask has a cover sealed onto the flask by means of a rubber collar. The cover has corres-ponding pipe connections for pouring helium, removing vapor, fastening a test object, etc.

To perform experiments in a magnetic field, the bottom part of both flasks is narrowed so that the distance between magnet poles is minimum. Superconducting solenoids are submerged directly into the helium flask, thus not requiring narrowing.

The drawbacks of glass helium cryostats are associated with their low strength and the necessity to carry out frequent vacuum treatment of the thermal insulation space. This is due to gaseous helium diffusion through the glass. Although more stable, borosilicate and molybdenum glasses are most permeable to helium. Therefore, special "Monax" or Kimbell glass is used for helium Dewar flasks.

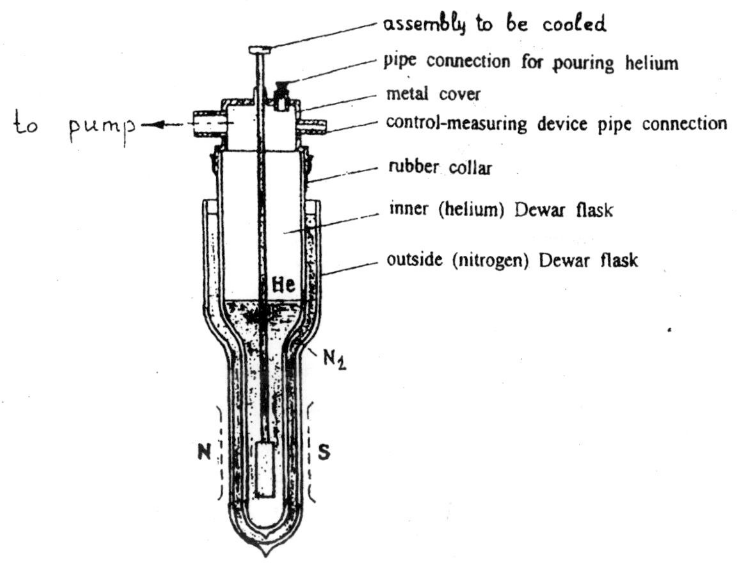

Fig. 16.2. Glass helium cryostat.

16.1.2. Helium cryostats

In metal helium cryostats, helium and nitrogen baths and cryostat body are manufactured from copper, stainless steel or aluminum alloys. On the side of the vacuum space, their surfaces are polished to reduce radiative heat transfer.

This type of cryostat has the following evaporation rates (Table 16.2).

Use of extra nitrogen cooling complicates a cryostats design and, in a number of cases, hampers studies. Liquid nitrogen use in helium cryostats may be excluded if evaporating helium cold is used to cool screens that reduce the heat flux from external wall emission. As the gaseous helium enthalpy is approximately 70 times higher than its heat of vaporization, even at a small evaporation rate the amount of the cold to be transferred by helium vapors is quite enough to

Table 16.2. Characteristics of metall helium cryostats.

Helium bath volume, l	Central tube diameter, mm	Liquid helium evaporation rate, $cm^3 \cdot h^{-1}$	Liquid helium flowrate, $l \cdot h^{-1}$
2.5 ÷ 3.5	36 ÷ 42	60 ÷ 90	0.1 ÷ 0.2
4 ÷ 6	48 ÷ 60	80 ÷ 140	0.2 ÷ 0.4

compensate the radiant heat flux.

The simplest helium cryostat design with no nitrogen cooling is shown in Fig. 16.3. The helium bath is suspended from the housing by a thin-wall pipe, through which the object to be cooled is placed into the cryostat, liquid helium is poured and evaporating helium is removed. Screens 4 and 5 in good heat contact with the pipe are mounted in the vacuum space between bath and housing. Thus, the helium vapor cold is transferred to the screens through the pipe. The number of screens and points of contact with the pipe must be chosen so as to provide cooling of the main screen up to approximately 70 ÷ 90 K. The larger the number of screens, the easier it is to meet this requirement; however, as practice shows, two screens are enough to obtain quite satisfactory results. To operate in a magnetic field, the bottom cryostat part is narrowed and has only one screen, thus allowing minimum clearance between inner ("helium") and outer ("heat") walls. In terms of evaporation rate of liquid helium, such cryostats have indices not worse than cryostats of similar size having extra nitrogen cooling.

In cases where studies must be made over a wide temperature range, or it is not desirable to place a test object in the fluid, sample cooling in cryostats is carried out by:

(1) obtaining gas at a corresponding temperature after liquid cryogent evaporation;

(2) coolant line having constant or periodic heat contact with liquid cryoagent;

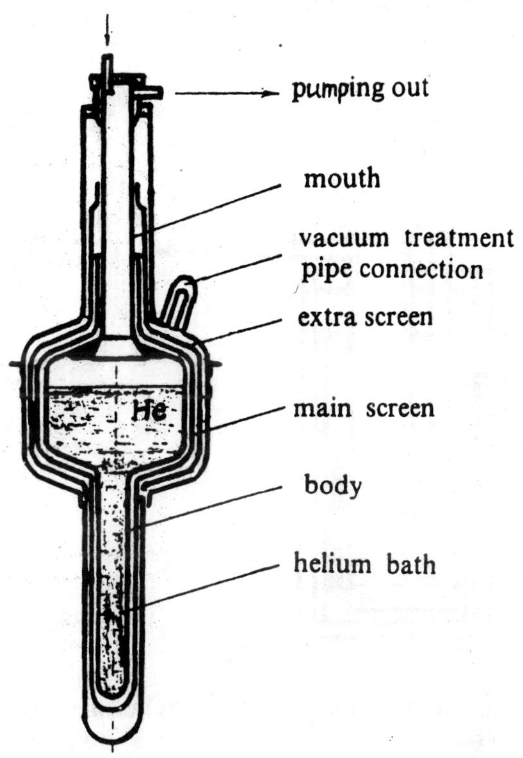

Fig. 16.3. Helium cryostat without nitrogen cooling.

(3) "heat transfer" gas removing the heat from the sample to a cryoagent bath.

In modern cryostats, an assigned sample temperature is usually maintained automatically.

Fig. 16.4 shows one of the widely used arrangements of a cryostat where an objects temperature is regulated by heated-gas blowing (Swenson's method). Liquid helium flows from the cryostat bath to a heat excanger where it evaporates and, after heating to the required temperature, is supplied to the working chamber containing the object to be studied. The temperature is regulated by a valve and a heater mounted on the outer surface of the heat exchanger. Temperatures above 80 K are achieved by filling the bath with liquid nitrogen instead of helium.

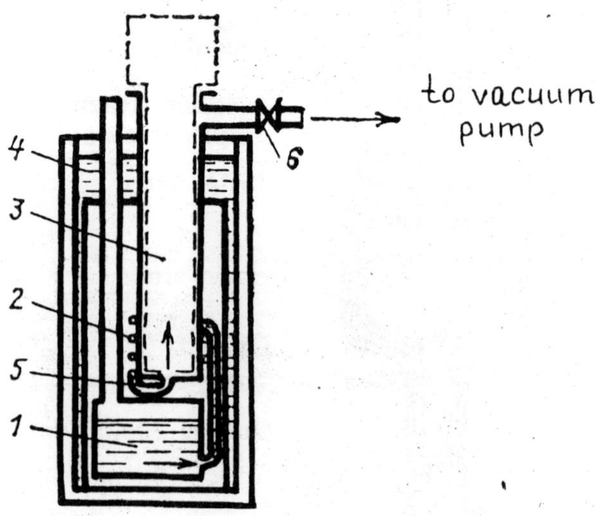

Fig. 16.4. Diagram of a temperature-controlled cryostat: 1, liquid helium bath; 2, heat exchanger; 3, working chamber; 4, nitrogen bath; 5, heater; 6, valve to regulate gas flow.

Cryostats used for optical and spectral studies are equipped with optical windows on the outer housing and sometimes also on the low-temperature working bath at helium temperatures.

Connecting optical windows to the metal chamber under helium temperatures is one of the problems encountered in manufacturing optical cryostats: it must be a vacuum-seal and must compensate any temperature deformations arising. The sealing of a pipe connection is made by epoxy resins or by annular gaskets made of indium and other soft metals or of non-metal materials, e.g. rubber film. A quite different method may also be used: a glass disc face serving as a window is metallized electrolytically, and then soft-soldered into a thin-wall copper cylinder.

For superconducting solenoids, it is most convenient to use cylindrical helium cryostats, with a large-diameter outlet mouth for free mounting and dismounting of magnet assemblies. Very often, the working volume of the magnetic field of a solenoid is also required to be at normal temperature.

Fig. 16.5 shows diagrams of several types of cylindrical helium cryostats for superconducting solenoids. The liquid helium flow-rate in such cryostats depends mainly on three factors: heat flux due to heat conduction of outlet mouth walls, cryostat cover radiation, and heat flux through electric inlets to the solenoid. Wall heat fluxes may be partially compensated by the cold of evaporating helium. For this, it is desirable to direct most of it along the wall but not over the mouth cross-section. Cover radiation may be also decreased by mounting screens in the outlet mouth, cooled with gas or liquid nitrogen.

Heat supply via current leads may be reduced if the latter are made of appropriate materials (depending on

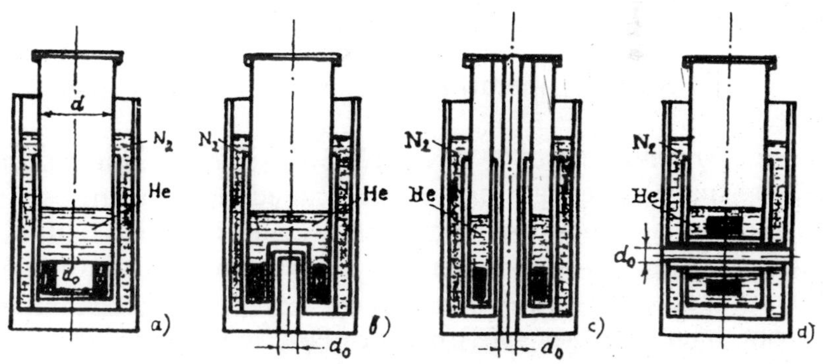

Fig. 16.5. Basic types of cylindrical cryostats for superconducting solenoids: a, standard type; b, c, with vertical working hole at normal temperature; d, with horizontal working hole at normal temperature.

operating conditions), have optimal sizes, and are cooled with evaporating helium. In a number of cases, these must be disconnected from the solenoid after it is energized.

16.1.3. Cryostats for outer-space research

Outer-space flight conditions (weightlessness, variable accelerations, high dynamic loads, etc.) prohibit the use of ordinary cryostats. Cryostats designed for space-craft have special features.

These are:

(1) mounting a device to hinder liquid cryoagent ejection from the working volume to outlet mouth of a cryostat under conditions of weightlessness and sign-variable accelerations. Such devices are called phase separators;

(2) making an outlet mouth design so as to eliminate (or to minimize) heat supply to the liquid cryoagent due to vapor convection that arises under real flight conditions;

(3) setting the regulator to maintain the pressure automatically above the liquid cryoagent accompanied by evaporating cryoagent exhaust into outer space;

(4) sufficient strength to endure all overloads and vibrations;

(5) elevated operating reliability for limited mass and size.

Another trend in developing cryostating systems for outer space investigations is associated with using solid cryoagent sublimation. For temperatures above 10 K, rocket-borne solid cryoagent cryostats have a number of advantages over liquid ones. They require no phase separators, have a longer safe operation and high reliability. The drawback of sublimation cryostats is that it is impossible to achieve and maintain helium temperatures in them.

As regards the number of solid cryoagents used, sublimation cryostats may be one- or two-component. Solid cryoagents are obtained by self-freezing of cryoproducts (pump-out of their vapors) and by heat exchange with a colder auxiliary cryoagent.

16.2. Low-temperature bubble chambers

The bubble chamber is one of the main pieces of equipment used in studies of high-energy particles.

The operation principle of a bubble chamber is the following. An accelerator particle beam is passed through a metastable fluid. Particles along their path generate electrons which, on decelerating, release energy giving rise to the onset and growth of nuclei up to a critical size; they then grow spontaneously. Thus, the paths of the primary particles that entered the accelerator chamber, and those of the secondary ones formed by nuclear reactions of the primary particles, together with the chamber substance, are seen as bubble chain tracks. Tracks are photographed when bubbles have grown to sizes corresponding to the resolution of the photographing system. Bubble chambers are placed in a magnetic field that deviates charged particle beams: positive particles in one direction and negative in another. By track curvature radius, magnetic induction and charge it is possible to determine the particle pulse.

At the same time, the substance filling a chamber serves, as a target. The simpler the substance composition, the easier is the analysis of nuclear reaction results. In particular, good results are obtained for reactions occurring on protons, or hydrogen nuclei. To study nuclear neutron reactions, hydrogen chambers are usually filled with deuterium. To work with deuterium, the chamber and its gas systems must have a higher pressure and must possess good sealing.

Propane is another working substance for bubble chambers. Propane contains carbon and hydrogen, thereby complicating irradiation data processing, while at the same time, in propane chambers (as in other heavy-liquid chambers) γ-quanta are recorded while in hydrogen chambers their path was not observed because of the long radiation wavelength, or a quantity that characterizes a photon path in a substance before conversion into an electron-positron pair. Hydrogen chambers are more complex than heavy-liquid ones as the need to work with cryotemperatures arises. Recently, several large hydrogen chambers have been redesigned so that in them hydrogen or deuterium might serve as the track-sensitive target, while recording secondary particles might be done by a neon-hydrogen mixture which possesses a small radiation wavelength and is therefore suitable for recording γ-quanta. Liquid helium chambers have not found wide use.

The optical layout of a hydrogen chamber, and hence its design, are to a considerable extent determined by the fact that the intensity of the light scattered on a bubble charply decreases with increase of angle between the directions of illumination and photography. In practice, it is impossible to photograph at an angle of 90^0 to the illumination direction. Trans-illumination photography is very simple: an lluminator is placed on the one side of the chamber, and photographic cameras on the other. For such a system, the chamber must have two windows allowing observation of the whole chamber volume in a plane parallel to the particle beam axis. However, as it is rather difficult to achieve good sealing between the window and chamber casing, most chambers are provided with only one large horizontal window. In this case, the illuminator and cameras are located on one side of the chamber casing. The opposite side (bottom) of the chamber is equipped with a reflection system in the form of a grating. The light scattered by the tracks along its path from the illuminator to the grating undergoes total internal reflection in the grating or is absorbed.

There is one more variant of an optical system when several small windows are left for cameras and illuminator. In this case, only some part of the chamber volume is used for recording tracks. Wide-angle objectives are utilized to decrease the parasitic chamber volume.

To reconstruct the track geometry in space, photography is performed stereoscopically using two or more cameras. The camera objectives are open permanently while the chamber is irradiated. To photograph, a pulse light flash is made in several milliseconds after the accelerator particle beam has passed. The tape mechanisms are engaged to prepare the camera for the next picture. Exposed film cassettes are replaced by new ones when needed.

The metastable state needed to form tracks is achieved by a pressure varying system. For the period between expansions, the chamber fluid is kept at a temperature and pressure above the normal boiling point (for fluid hydrogen, the temperature range is 26 ÷ 29 K at an absolute pressure 0,4 to 0,8 MPa). Then, at the required moment (some time before passing the particle beam), the chamber pressure is quickly relived, and for some time the fluid is metastable; then the pressure is again elevated, not allowing fluid boiling within the entire volume. Liquid hydrogen possesses marked compressibility.

Piston or bellows devices are most widely used for varying chamber pressure.

Glass sealing in a metal casing is rather difficult problem since the glass and metal of a chamber have different linear expansion (compression coefficients). While cooling a chamber the glass must slide on the seal. At the same time, the seal must be kept between liquid-hydrogen and high-vacuum spaces. Large oval-shaped pieses of glass are sealed by inflatable gaskets. Indium possesses plasticity at low temperatures and therefore serves as a gasket. The thermal insulation system for the majority of chambers is high-vacuum or a vacuum flask.

The thermostating system must maintain a chamber fluid at a certain temperature, accurate to 0.01 ÷ 0.05 K.

In hydrogen chamber, the pressure-varying system releases heat, approx. 0.3 ÷ 1.2 $J \cdot (l \cdot cycle)^{-1}$. Morecver, the thermostating system must remove environmental heat fluixes. Usually, thermostating is carried out by hydrogen circulation via a system of pipes, whose connection depends on chamber desing. In the majority of cases, cooling liquid hydrogen enters from the hydrogen liquefier located near the chamber. Evaporating cold hydrogen is returned to the liquefier which operates under refrigerated conditions. In this case, the refrigerating capacity of a cryogenic plant is doubled, compared to the liquefaction conditions (no cold vapor return). There are several chambers where for the sake of safety, a separate helium cycle is used for hydrogen recondensation.

Stabilizing a chamber temperature is regulated by liquid hydrogen flowing through cooling coils. Usually, chamber pressure is in more than several hundred fractions of a megapascal over equilibrium at a corresponding temperature. In the majority of cases, a constant of hydrogen is kept in a chamber and the pressure changes as a function of temperature. In cases where the chamber pressure and temperature are regulated independently, the chamber is filled with hydrogen via a condenser at a lowered pressure. At an elevated pressure, the hydrogen excess is extracted by a pressure regulator.

Recently, the leading laboratories of the world have started using many-meter bubble chambers: the European bubble chamber (CERN) 3.7 m in dia; the 15-foot chamber at the Fermi laboratory (USA); the French 4.5 m chamber.

Use of superconducting magnets enables magnetic induction to be raised whilst reducing energy consumption. To speed up data acquisition, modern chambers use a many-cycle operating regime, i.e. several cycles to relieve and elevate the pressure for one cycle of accelerator operation.

16.3. Loops for studies in nuclear reactors at cryotemperatures

A large number of physical experiments need a slow neutron flux. Neutrons with an energy below about 0.005 eV and wavelength of 0.4 nm or more are considered slow (or "cold"). The neutron flux developing in a reactor at a temperature of approx. 50 °C and at a thermal equilibrium with a moderator contains less than 2% "cold" neutrons. To obtain a fairly dense flux of "cold" neutrons, the latter must be decelerated at the low temperatuire. To achieve this, the low-temperature moderator reservoir is placed along the beam path. This is usually liquid hydrogen or deuterium or a mixture. Sometimes liquid helium is used as a moderator. This cryosystem is frequently called a "loop". Fig. 16.6 shows a schematic of a "loop" for obtaining "cold" neutrons by moderating them in liquid hydrogen which condenses on heat exchange with the gaseous helium flow.

Apart from obtaining a "cold" neutron beam, "loops" are used to study the nuclear radiation effect on a substance. Studying defects and varying the physical properties that arise in irradiated solids is difficult at normal temperature because of the defect instability due to thermal oscillations. It is therefore necessary both to irradiate and study solids at low temperatures, so that the defects seem "frozen" and can be examined in a the state they arose under irradiation.

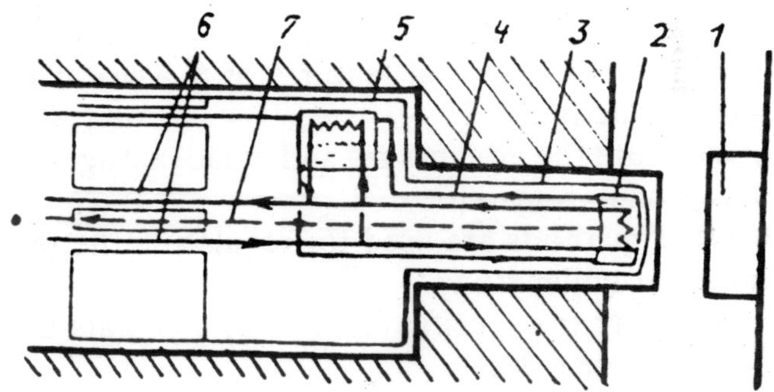

Fig. 16.6. Schematic of a "loop" for obtaining the flux of "cold" neutrons moderated in liquid hydrogen: 1, reactor core; 2, liquid hydrogen vessel; 3, screen; 4, evaporating hydrogen pipeline; 5, hydrogen condenser; 6, circulating helium pipeline; 7, "cold" neutron flux.

Cases are known where defects already vary at liquid nitrogen temperature (77 K). At present, cryosystems are being developed that allow a substance to be irradiated and studied at three characteristic temperature levels: 77 K and above, 20 K and approximately 4 K.

Designing "loops" to be used for studies in nuclear reactors at cryogenic temperatures must take account of the following facts:

(1) difficult access, with required equipment, to the reactor core; for better safety inlet and outlet cryoagent pipes must be made very long (dozens of meters);

(2) since the neutron flux is always accompanied by γ-radiation, in the sample material, liquid cryoagent and cryostat walls a considerable amount of heat is released (from tens to several hundreds of watts or even more) and must be removed by circulating cryoagent. It is therefore important to choose the right material for pipelines and cryostats. Of most interest are aluminum and magnesium (15 h), and zirconium (17 h) because of their low half-decay time.

Questions

1. What are the special features of a cryostat for studying superconductivity?
2. What are the advantages and disadvantages of glass cryostats as used in research?
3. What methods are used for regulating the temperature of an object under study?
4. What are the specific features of cryostats as applied on board a space vehicle?
5. What is the principle of operation of a low-temperature bubble chamber?
6. What working bodies are used in bubble chambers?
7. What is the design of a "loop" for research in a nuclear reactor at cryotemperatures?